Illustrated Guide : English for Arithmetic

図解
子供にも教えたい
算数の英語
豊富な用語と用例

銀林 浩 ● 銀林 純

日興企画

はじめに

●小学校教育の国際化

　国際化のかけ声のもと，最近では公立の小学校での英語教育の試行なども始まりました。そこで，算数（だけでなく他の科目もですが）を英語で学ぶ機会もますます増えていくことでしょう。

　しかし，海外勤務の親に連れられて小さい頃から英語国に住み，現地の小学校に通ったり，日本国内でもインターナショナル・スクールと呼ばれる学校に在籍したりしていれば，そうした経験も可能ですが，そうでないかぎりは不可能でしょう。今日の日本では，学校で学ぶ子どもはいうまでもなく，その親やまわりの大人にとっても，英語で算数をするというのはまず初めての経験であろうと思われます。

　そこで，この本は，算数の教科書に登場するさまざまな用語や語句，規則や文章題を英語でどう表現するか，その方法について楽しく学べるようにやさしく，ていねいに解説しています。算数の解説もさることながら英語表現を重視して書かれています。

●算数の英語表現を単元ごとに解説

　この本では，小学校で学ぶ算数を大きく「いろいろな数」「数式と計算」「量と単位」「平面図形と空間図形」「面積と体積」「表とグラフ」「比・比例と場合の数」という7つの章に分け，さらにその中をいくつかの単元に整理してあります。

　この章や単元の分け方とその順序はかならずしも小学校で習う順番ではありませんが，教科書を念頭に分野別になっていますので，親あるいは高学年の子にとっては学びやすいでしょう。低学

年の子が学ぶときには，適切な学習範囲を親や兄姉がガイドしてあげると良いでしょう。また，「年齢」や「通貨」など算数の単元としては取りあげにくかったテーマの英語表現をいくつかコラムで紹介しています。単元の順番とは無関係なので，いつでも休憩がてら読んでみてください。

●基本用語・語句・規則・出題例を豊富に紹介

各単元には，基本的な用語や語句の英語表現と，それらを用いた規則や出題例を豊富に載せています。基本的な用語ほど，その英語表現は（大人の）日本人には意外に思えたり，あるいは驚きに感じたりするものがあります。ですから基本用語だけ読み進んでもじゅうぶん勉強になり，おもしろいでしょう。また，例文はできるだけやさしい文例から少し複雑な文例までを順番に並べてあり，無理なく読み進められるように工夫してあります。

●イラストや図解による視覚的な説明

「算数にチャレンジしてみたい」，最近，そんな大人が増えているようですが，算数の用語や規則などが簡単に，しかもはやく理解できるようにイラストや図解をたくさん使って視覚的に説明しています。ですからイメージが描きやすく，大人の復習にはもちろん，子どもと一緒の学習にもおすすめです。

●練習問題にチャレンジ

各章に「チャレンジ・テスト」を載せています。少し長めですが，復習もかねて英文での文章題に挑戦してみてください。解答もできるだけ英語でトライしてください。解答例を巻末につけてあります。

●英語で算数を学ぶ利点

　ところで，ただでさえ厄介な英語でわざわざ算数を学ぶことに何の利点があるのかと不審に思われるかもしれません。じつは，英語で算数を学ぶことは，算数にとっても英語学習にとっても具合が良いのです。

　数学は論理的で，しかも抽象的な学問です。それを反映して算数もロジカルにできていて，しかも英語での言いまわし（公式）も決まっていますから子どもでもすぐ慣れます。それに小学校段階では場面設定が比較的単純なので，必要な語彙は少なくてすみます。これらは外国語の学習にとって都合の良い点といえます。つまり，算数は外国語学習のすぐれた材料の一つなのです。

　一方，今日の数式や記号の多くは16世紀にイギリスやヨーロッパで作られたものです。たとえば，「2＋3＝5」を日本の学校では「2たす3は5」と読ませていますが，これはもともと英語のラフな言い方なら「2 and 3 is 5.」で，「and（と／たす）」にあたるラテン語「et」を変形して「＋」にし，「is」を平行線「＝」に変えただけのものなのです。つまり，数式は英語の文章そのものともいえます。英語で算数を学ぶことは算数の起源（ルーツ）を探ることにもなるわけで，算数の理解を深められることでしょう。

　最後に，この本を通して「算数って，おもしろい」「へえ～，英語ではこんなふうに表現するのか」と楽しんだり，驚いたりしていただけたら，こんなにうれしいことはありません。

2006年1月

　　　　　　　　　　　　　　　　　　　　銀林浩・銀林純

目次　図解・子供にも教えたい算数の英語●豊富な用語と用例

- ●―はじめに ─────────────────── 3
- ●―ガイダンス ────────────────── 10

第1章　いろいろな数
◆Various Numbers

1. 個数の数（基数）◆Cardinal Numbers ──────── 12
2. 順序の数（序数）◆Ordinal Numbers ───────── 16
3. 位（桁）◆Places ────────────────── 18
4. 偶数と奇数◆Even Numbers and Odd Numbers ──── 20
5. 倍数と約数◆Multiples and Factors ───────── 22
6. 概数と概算（およその数）
   ◆Approximation and Estimation ────────── 25
7. 数直線上の数◆Various Numbers on Number Lines ── 27
8. 小数◆Decimals ─────────────────── 30
9. 分数とわり算◆Fractions and Division ─────── 33
- ●―チャレンジ・テスト①◆Challenge Test ① ───── 37

第2章　数式と計算
◆Expressions and Operations

1. たし算（整数・分数・小数）◆Addition ─────── 44
2. ひき算（整数・分数・小数）◆Subtraction ───── 47
3. かけ算九九◆Multiplication Tables ──────── 51
4. かけ算（整数・分数・小数）◆Multiplication ──── 53
5. わり算（整数・分数・小数）◆Division ─────── 56

6．四則混合算◆Mixed Operations ──────59
　7．文字と式◆Variables and Expressions ──────62
　●──チャレンジ・テスト②◆Challenge Test ② ──────66

第3章　数量と単位
◆Numerical Quantities and Units

　1．時間と時刻◆Time and Duration ──────72
　2．長さ◆Length ──────75
　3．かさ（液量・容積）◆Liquid Measurement ──────78
　4．重さ◆Weight ──────82
　5．平均と密度と速さ◆Average, Density and Speed ──────85
　●──チャレンジ・テスト③◆Challenge Test ③ ──────88

第4章　平面図形と空間図形
◆Plane Figures and Solid Figures

　1．点と線（垂直・平行）◆Points and Lines ──────94
　2．角と角度◆Angles and their Measure ──────97
　3．3角形◆Triangles ──────100
　4．4角形（正方形・長方形・台形・平行4辺形・ひし形）
　　　◆Quadrilaterals ──────103
　5．多角形◆Polygons ──────105
　6．合同・相似と角
　　　◆Congruence, Similarity and Angles ──────108
　7．線対称と点対称
　　　◆Line Symmetry and Point Symmetry ──────112
　8．円と円周◆Circles and Circumference ──────115

9．立方体・直方体と展開図
　　　◆Cubes, Cuboids and Nets ——————————117
10．柱と錐と球◆Prisms, Pyramids and Spheres ——————121
●──チャレンジ・テスト④◆Challenge Test ④ ——————126

第5章 面積と体積
◆Areas and Volumes

1．面積・体積・容積と単位
　　　◆Areas, Volumes, Capacities and Units ——————136
2．3角形・4角形・多角形と円の面積
　　　◆Area of Triangles, Quadrilaterals, Polygons and Circles —140
3．方体・柱・錐・球の表面積
　　　◆Surface Areas of Cuboids, Prisms, Pyramids and Spheres—142
4．方体・柱・錐・球の体積
　　　◆Volume of Cuboids, Prisms, Pyramids and Spheres ——146
●──チャレンジ・テスト⑤◆Challenge Test ⑤ ——————150

第6章 表とグラフ
◆Tables and Graphs

1．表と棒グラフ・折れ線グラフ
　　　◆Table, Bar Graph and Line Graph ——————————154
2．百分率と円グラフ・帯グラフ
　　　◆Percent, Circle Graph and Bar Graph ——————————156
3．平均・分布・延べと柱状グラフ
　　　◆Average, Distribution, Total and Histogram ——————158
●──チャレンジ・テスト⑥◆Challenge Test ⑥ ——————161

第7章 比・比例と場合の数
◆Ratios, Proportion and Number of Cases

1．比と比例◆Ratios and Proportion ──────────166
2．拡大と縮小（相似）
　　◆Enlargement and Reduction (Similarity)──────170
3．場合の数（順列・組合せ・確率）
　　◆Permutations, Combinations and Probability ────174
●─チャレンジ・テスト⑦◆Challenge Test ⑦ ─────177

●─解説◆数の英語表現，ここがポイント────────181

●─Tea Room
①年齢・○歳代・年数── 42　②年月日・年代・世紀──134
③書籍・雑誌・新聞───164　④通貨 ─────────180

●─チャレンジ・テストの解答 ───────────────187

装丁＋本扉＋中扉 …………………………高橋すさ子
本文印字＋レイアウト＋図版 ………………m2.design
本文イラスト ………………………………アイコウ カオリ

ガイダンス

① − 各章は独立して書かれていますので，どの章から読み始めてもかまいませんし，興味のある章だけを読むこともできます。また，各章の各単元は大きくわけて基礎的な用語や語句（枠囲み）と，算数の規則や典型的な文章題（枠囲み以外）の二つの部分から成り立っています。

② − いろいろな数の書き方と読み方については巻末にまとめて解説してありますので，整数・小数・分数などおたがいの関連を考えながら学習していくことをおすすめします。コラムも数字表現のいろいろがテーマです。あわせてお読みください。

③ − 用語が可算名詞の場合，文脈上とくに複数になる場合を除き，基本的に単数形で説明しています。

④ − 数は算用数字と英語表現を併記しています。

⑤ − 省略や置き換え可能な表現部分についてはつぎのように表記しています。

   [A] ………………Aは省略可

   A（B）…………Aの代わりにBでもよい。

⑥ − 基本的にアメリカ英語での表現を紹介していますが，とくにアメリカ英語とイギリス英語を区別するときは【米】と【英】がつけてあります。

⑦ − 発音を表わす特別な書き方としてつぎがあります。

   oh ………………0（zero）の読み方の一つで「オウ」。

第1章 いろいろな数

Various Numbers

① 個数の数（基数）
Cardinal Numbers

＊基数の読み方と書き方については巻末の「解説」をご覧ください。

基数の名前

1	one	2	two
3	three	4	four
5	five	6	six
7	seven	8	eight
9	nine	10	ten
11	eleven	12	twelve
13	thirteen	14	fourteen
15	fifteen	16	sixteen
17	seventeen	18	eighteen
19	nineteen	20	twenty
30	thirty	40	forty
50	fifty	60	sixty
70	seventy	80	eighty
90	ninety	100	one hundred

大きな数

1,000 （1千） …………………one thousand
10,000 （1万） …………………ten thousand
100,000 （10万） …………………one hundred thousand
1,000,000 （100万） …………………one million

10,000,000（1,000万）……… ten million
100,000,000（1億）……………one hundred million
1,000,000,000（10億） ……… one billion【米】
　　　　　　　　　　　　　 …one thousand million【英】
1,000,000,000,000（1兆） ……one trillion【米】
　　　　　　　　　　　　　 …one billion【英】

数の読み方

84　　………………………eighty-four
106　………………………one hundred [and] six
234　………………………two hundred [and] thirty- four
504　………………………five hundred [and] four
2,009 ………………………two thousand [and] nine
3,087 ………………………three thousand [and] eighty-seven
70,004（7万4）…………seventy thousand and four
903,472（90万3472）
　………nine hundred and three thousand, four hundred
　　　　[and] seventy-two
6,498,735（649万8735）
　………six million, four hundred [and] ninety-eight
　　　　thousand, seven hundred [and] thirty-five
7,378,905,423（73億7890万5423）
　………seven billion, three hundred [and] seventy-eight
　　　　million, nine hundred and five thousand, four
　　　　hundred [and] twenty-three【米】
　　　　seven thousand, three hundred [and] seventy-
　　　　eight million, nine hundred and five thousand,
　　　　four hundred [and] twenty-three【英】

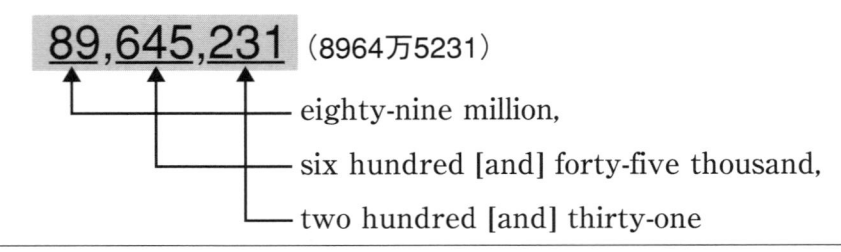

89,645,231 (8964万5231)
- eighty-nine million,
- six hundred [and] forty-five thousand,
- two hundred [and] thirty-one

大きな数の読み方／米と英の違い

3,657,904,562,967 (3兆6579億456万2967)

【米】
- three trillion,
- six hundred [and] fifty-seven billion,
- nine hundred and four million,
- five hundred [and] sixty-two thousand,
- nine hundred [and] sixty-seven 【米】

3,657,904,562,967 (3兆6579億456万2967)

【英】
- three billion,
- six hundred [and] fifty-seven thousand,
- nine hundred and four million,
- five hundred [and] sixty-two thousand,
- nine hundred [and] sixty-seven 【英】

＊－millionより大きい数の読み方・書き方は【米】と【英】では違います。【米】ではmillionの先も3桁ずつ区切り、【英】では6桁ずつ区切ってbillion, trillion……と進みます。

●100を3個，10を6個，1を9個あわせた数は369です。

The number you get when three 100's (hundreds) and six 10's (tens) are added to nine 1's (ones) is 369 (three hundred [and] sixty-nine).

●5427は1000を5個，100を4個，10を2個，1を7個あわせた数のことです。

5,427 (Five thousand, four hundred [and] twenty-seven) is the number that five 1000's (thousands), four 100's (hundreds), and two 10's (tens) are added to seven 1's (ones).

●下の数直線を見てAとBはそれぞれいくつでしょう。

Which numbers are A and B respectively on the following number line?

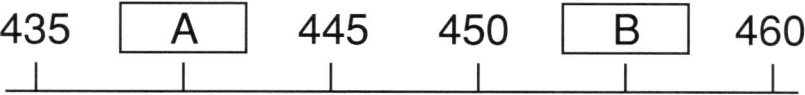

●日本の人口はおおよそ1億2千5百万人です。

The population of Japan is approximately (about) 125,000,000 (one hundred [and] twenty-five million).

●川崎市では土地を23億円で買い，建物を8億円かけてスポーツ公園をつくりました。あわせていくらかかったでしょう。

Kawasaki City built a sports park by buying land with 2,300,000,000 (two billion, three hundred million【米】／ two thousand, three hundred million【英】) yen and constructing a building with 800,000,000 (eight hundred million) yen. How much did it cost in total?

❷ 順序の数（序数）
Ordinal Numbers

＊序数の読み方と書き方については巻末の「解説」をご覧ください。

序数の名前と読み方

1番目 ……first	2番目 ……second
3番目 ……third	4番目 ……fourth
5番目 ……fifth	6番目 ……sixth
7番目 ……seventh	8番目 ……eighth
9番目 ……ninth	10番目 ……tenth
11番目 ……eleventh	12番目 ……twelfth
13番目 ……thirteenth	14番目 ……fourteenth
15番目 ……fifteenth	16番目 ……sixteenth
17番目 ……seventeenth	18番目 ……eighteenth
19番目 ……nineteenth	20番目 ……twentieth
30番目 ……thirtieth	40番目 ……fortieth
50番目 ……fiftieth	60番目 ……sixtieth
70番目 ……seventieth	80番目 ……eightieth
90番目 ……ninetieth	100番目 ……one hundredth
1000番目 ……one thousandth	10000番目 ……ten thousandth

＊　　　＊　　　＊

67番目 ……sixty-seventh
111番目 ……one hundred eleventh
341番目 ……three hundred forty-first
3456番目 ……three thousand four hundred fifty-sixth

●東京駅から7つ目はなんという駅ですか。

　Which station is the 7th (seventh) from Tokyo Station?

●明君はマラソンで48番になりました。

　Akira was the 48th (forty-eighth) in the marathon.

●太郎君の席は前から3番目です。

　Taro's seat is the 3rd (third) from the front.

●ここに8人並んでいます。前から3番目は後から何番目ですか。

　8 (Eight) people are standing in line. In which position from the end does the 3rd (third) person stand?

●競技で，太郎君は2着で，次郎君はさらに3番おくれました。次郎君は何着ですか。

　In the competition, Taro was the 2nd (second) and Jiro was the 3rd (third) from Taro. What was Jiro's result?

③ 位 (桁)

Places

位の名前

位	place
位取り	position of figure
1の位の数字	the 1's (ones) digit
10の位の数字	the 10's (tens) digit
100の位の数字	the 100's (hundred's) digit

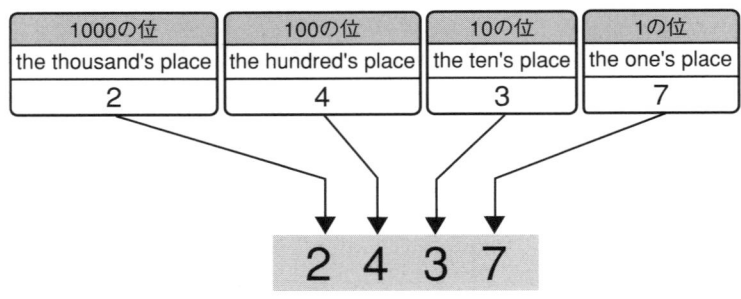

桁数

1桁の数…single figures……3, 5, 7, 4, 6

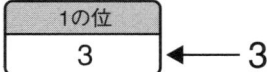

2桁の数…double figures……23, 45, 87, 69

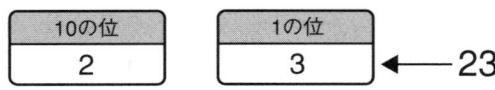

3桁の数…three-figure numbers……234, 456, 753, 446

上から2桁目の数…the 2nd(second) place digit from the top

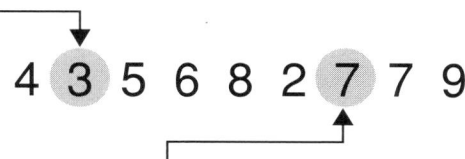

下から3桁目の数…the 3rd(third) place digit from the bottom

＊一一般にn桁の数はn-figure numbersといいますが，1桁と2桁の場合は個別の呼び方をします。

○進法

10進法……………………………………the decimal system
2進法……………………………………the binary system

●457は3桁の数字です。

457 (Four hundred [and] fifty-seven) is a three-figure number.

●764の10の位はいくつですか。

What is the tens place digit of 764 (seven hundred [and] sixty-four)?

●コンピュータは2進法でできています。

Computers are based on the binary system.

第1章●いろいろな数／Various Numbers

④ 偶数と奇数
Even Numbers and Odd Numbers

整数の名前

自然数	natural (counting, whole) number / non-negative (positive) integer
整数	integer
素数	prime number
偶数	even number
奇数	odd number

● 0は自然数ではありません。

0 (Zero) is not a natural number.

● 1は素数ではありません。

1 (One) is not a prime number.

● 2の倍数を偶数といいます。

The multiples of 2 (two) are called even numbers.

● 偶数でない整数を奇数といいます。

An integer that is not an even number is called an odd number.

● 偶数より1大きい数は奇数になります。

The number which is one more than an even number is an odd number.

●2でわりきれる整数を偶数といい，余りが1になる整数を奇数といいます。

An integer is called an even number if it is divisible by 2 (two), and is called an odd number if the remainder becomes 1 (one) when divided by 2 (two).

●整数では奇数と偶数は交互にでてきます。0は偶数にふくめます。

In integers odd numbers alternate with even numbers. 0 (Zero) is included in even numbers.

●8，13，35，67，42，81，246を偶数と奇数にわけましょう。

Divide the numbers 8 (eight), 13 (thirteen), 35 (thirty-five), 67 (sixty-seven), 42 (forty-two), 81 (eighty-one), and 246 (two hundred [and] forty-six) into even numbers and odd numbers.

●奇数と奇数をたすと偶数，奇数と偶数をたすと奇数になります。

The sum of two odd numbers is an even number, and the sum of an odd number and an even number is an odd number.

●奇数は，奇数をかけると奇数，偶数をかけると偶数になります。

An odd number becomes another odd number when multiplied by yet another odd number, and becomes an even number when multiplied by another even number.

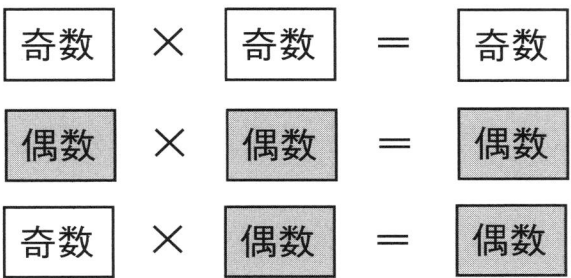

❺ 倍数と約数
Multiples and Factors

倍数の読み方

倍数 …………… multiple
公倍数 ………… common multiple
最小公倍数 …… the least common multiple (LCM) 【米】
the lowest common multiple (LCM) 【英】

    *   *   *

3の倍数………… multiples of 3 (three)
4の倍数………… multiples of 4 (four)
13の倍数 ……… multiples of 13 (thirteen)
4と9の公倍数 … common multiples of 4 (four) and 9 (nine)
21と32の公倍数
 …… common multiples of 21 (twenty-one) and 32 (thirty-two)

約数の読み方

約数 …………… factor
公約数 ………… common factor (common divisor)
最大公約数 …… the greatest common factor (GCF) 【米】
      the greatest common divisor (GCD) 【米・英】

    *   *   *

9の約数 ……… factors of 9 (nine)
12の約数 ……… factors of 12 (twelve)
24の約数 ……… factors of 24 (twenty-four)

4と8の公約数
　……common factors (divisors) of 4 (four) and 8 (eight)
12と32の公約数
　……common factors (divisors) of 12 (twelve) and 32 (thirty-two)

●ある数に整数をかけてできる数を元の数の倍数といいます。

The result we get when multiplying a number with an integer is called a multiple of the original number.

●2つ以上の整数に共通な倍数を公倍数といいます。

A multiple which is common to two or more integers is called a common multiple.

●いちばん小さい公倍数を最小公倍数（ＬＣＭ）といいます。

The smallest common multiple is called the lowest common multiple (LCM).【英】

●12の倍数を4つ書いてください。

Write four multiples of 12 (twelve).

12の倍数　⟶　12, 24, 36, 48, 60, 72, 84 ……

●3と5と7の公倍数を小さいほうから3ついってください。

Say three common multiples of 3 (three), 5 (five) and 7 (seven) starting with the smallest one.

●6と8の最小公倍数を求めましょう。

Let's find the least (lowest) common multiple of 6 (six) and 8 (eight).

6の倍数 ⟶ 6, 12, 18, ㉔, 30, 36, 42, 48 ……

8の倍数 ⟶ 8, 16, ㉔, 32, 40, 48, 56, 64 ……

●ある整数をわりきることのできる数を元の整数の約数といいます。

A number by which an integer is divisible is called a factor of the original integer.

●いちばん大きい公約数を最大公約数といいます。

The largest common factor is called the greatest common factor.

●18の約数を見つけましょう。

Let's find the factors of 18 (eighteen).

●28と35の公約数はいくつですか。

What are the common factors (divisors) of 28 (twenty-eight) and 35 (thirty-five)?

28の約数 ⟶ ①, 2, 4, ⑦, 14, 28

35の約数 ⟶ ①, 5, ⑦, 35

●12と16の最大公約数は4です。

The greatest common factor (divisor) of 12 (twelve) and 16 (sixteen) is 4 (four).

12の約数 ⟶ 1, 2, 3, ④, 6, 12

16の約数 ⟶ 1, 2, ④, 8, 16

⑥ 概数と概算（およその数）
Approximation and Estimation

四捨五入・切り捨て・切り上げ

概数 …………………………………… approximation
有効数字 ……………………………… significant figures
概算 …………………………………… estimation

```
20  21  22  23  24  25  26  27  28  29  30
```

1の位を四捨五入する……round [off]
●―――30―――▶●
◀―――20―――○

1の位を切り捨てる………round down
●―――――――20―――――――○

1の位を切り上げる………round up
○―――――――30―――――――●

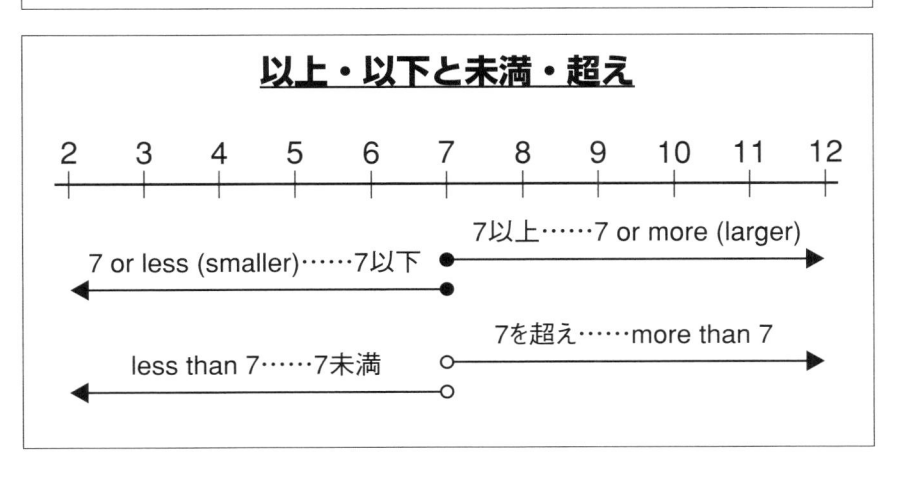

以上・以下と未満・超え

```
2   3   4   5   6   7   8   9   10  11  12
```

7以上……7 or more (larger)
7 or less (smaller)……7以下
7を超え……more than 7
less than 7……7未満

●およその数を概数といいます。

A rough number is called an approximation.

●概数を使った計算を概算といいます。

Calculation with approximate numbers is called estimation.

●概数は四捨五入，切り上げ，あるいは切り捨てで得られます。

An approximation can be obtained by rounding [off], rounding up, or rounding down.

●1の位を四捨五入する

round [off] to the nearest ten

●四捨五入して上から3桁までの概数にしましょう。

Let's get an approximation with three significant figures by rounding.

●3427の100の位以下を切り捨てると，3000になります。

3,427 (Three thousand, four hundred [and] twenty-seven) becomes 3,000 (three thousand) when it is rounded down to the nearest thousand.

●今日の動物園の入場者数は約23000人です。

The number of today's visitors to the zoo is approximately (about) 23,000 (twenty-three thousand) people.

●野球チームは9選手以上が必要です。

A baseball team needs 9 (nine) or more players.

❼ 数直線上の数
Various Numbers on Number Lines

大小の記号と式

不等号（＜　＞）……………………inequality sign
等号（＝）……………………………equality sign／equal sign
等式 ……………………………………equality (equation)
不等式 …………………………………inequality (inequation)
辺 ………………………………………side
右辺 ……………………………………right side／right-hand side
左辺 ……………………………………left side／left-hand side
両辺 ……………………………………both sides

大小関係

□＜△　「□は△より小さい」　　□ is less than △.

356＜385……356は385より小さい

356（Three hundred [and] fifty-six） is less than 385（three hundred [and] eighty-five）.

□＞△　「□は△より大きい」　　□ is greater than △.

427＞296……427は296より大きい

427（Four hundred [and] twenty-seven） is greater than 296（two hundred [and] ninety-six）.

数の大小

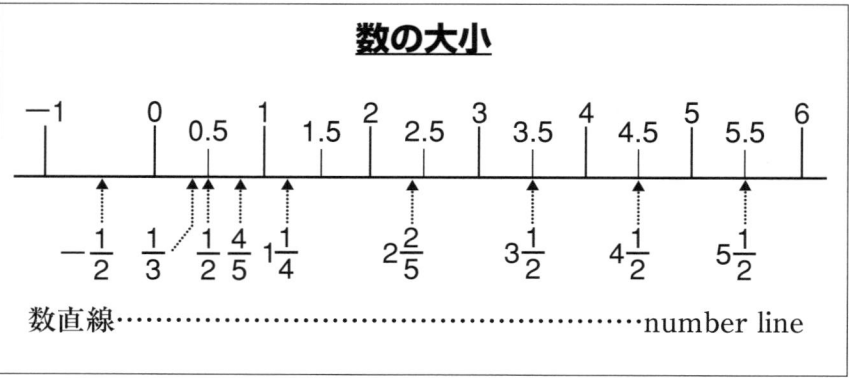

数直線 …………………………………………… number line

正負の数

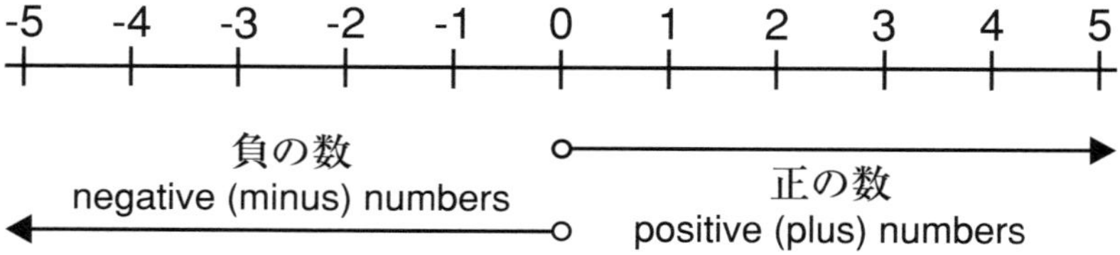

```
 -5  -4  -3  -2  -1   0   1   2   3   4   5
 ─┼───┼───┼───┼───┼───┼───┼───┼───┼───┼───┼─
```

負の数 / negative (minus) numbers
正の数 / positive (plus) numbers

正の整数 …………………………………… positive integer
負の整数 …………………………………… negative integer
＋（プラスの記号）……………………… the positive sign
－（マイナスの記号）…………………… the negative sign
同符号 ……………………………………… like (same) sign
異符号 ……………………………………… unlike (different) sign

　　　　　　　＊　　　＊　　　＊

＋25 …………………………………………… positive twenty-five
－8 ……………………………………………… negative eight
＋9 －（－7）……… positive nine minus negative seven
－8 ＋（＋6）……… negative eight plus positive six

●38と49はどちらが大きいですか？

Which of 38 (thirty-eight) and 49 (forty-nine) is greater?

●26と46の大小関係を不等式で表わしなさい。

Use an inequality to show which of 26 (twenty-six) and 46 (forty-six) is greater.

●0は正でも負でもありません。

0 (Zero) is neither positive nor negative.

●−3，8，0，−7，5を小さい順に並べなさい。

Put (List, Arrange) [the numbers] −3(negative three), 8(eight), 0(zero), −7(negative seven), and 5 (five) in order from least to greatest (in ascending order).

⑧ 小数
Decimals

＊小数の読み方と書き方については巻末の「解説」をご覧ください。

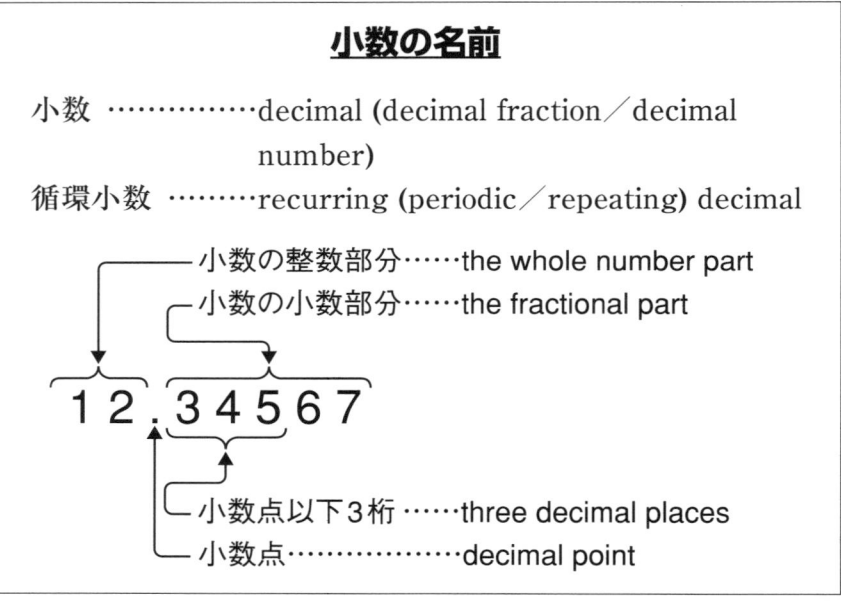

小数の名前

小数	decimal (decimal fraction／decimal number)
循環小数	recurring (periodic／repeating) decimal

- 小数の整数部分……the whole number part
- 小数の小数部分……the fractional part

1 2 . 3 4 5 6 7

- 小数点以下3桁……three decimal places
- 小数点……decimal point

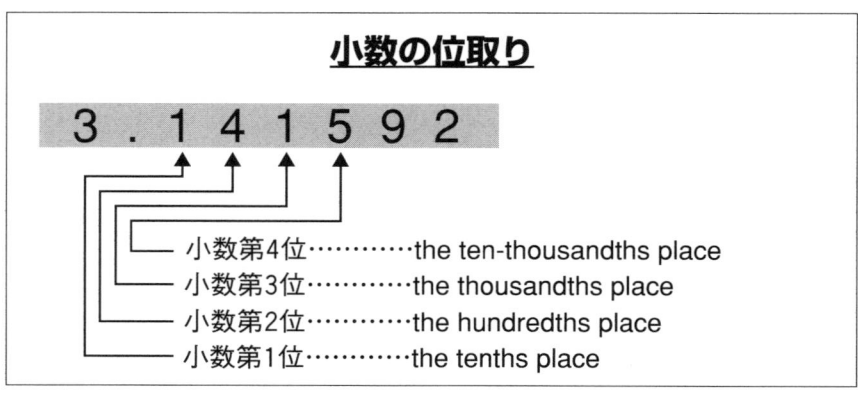

小数の位取り

3 . 1 4 1 5 9 2

- 小数第4位…………the ten-thousandths place
- 小数第3位…………the thousandths place
- 小数第2位…………the hundredths place
- 小数第1位…………the tenths place

小数の読み方

0.1 ……………one tenth
0.01 ……………one hundredth
0.001 ……………one thousandth
0.03 ……………three hundredths (zero point zero three)
0.47 ……………forty-seven hundredths
　　　　　　　(zero point four seven)
0.369 ……………three hundred sixty-nine thousandths
　　　　　　　(zero point three six nine)
2.345 ……………two and three hundred forty-five
　　　　　　　thousandths (two point three four five)
68.852 …………sixty-eight and eight hundred fifty-two
　　　　　　　thousandths
　　　　　　　(sixty-eight point eight five two)

●0.2を4つ集めた数は0.8です。

If we add four 0.2's (two tenths) we get a total of 0.8 (eight tenths).

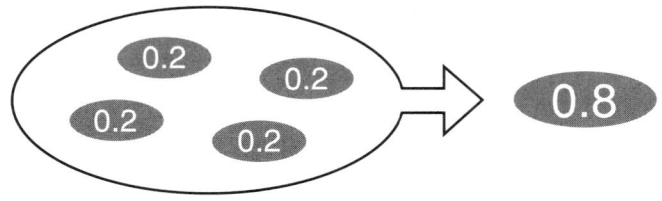

●320cmは3.2mです。

320 cm (Three hundred [and] twenty centimeters) is equal to 3.2 m (three point two meters).

● 4.67kgは何gでしょうか。

How many grams is 4.67 kg (four point six seven kilograms)?

● 0.83の10倍は8.3です。

Ten times 0.83 (zero point eight three) is 8.3 (eight point three).

$$0.83 \times 10 \Rightarrow 8.3$$

● 3.4869の小数第3位の数はいくつですか。

What is the 1,000ths (thousandths) place digit of 3.4869 (three point four eight six nine)?

● 0.7を分数で表わしてください。

Write 0.7 (zero point seven) as a [vulgar] fraction.

● 1.25を分数で表わすと，$1\frac{25}{100}$，約分すると，$1\frac{1}{4}$になります。

1.25 (one point two five) becomes $1\frac{25}{100}$ (one and twenty-five over one hundred) when written as a fraction, and $1\frac{1}{4}$ (one and one fourth) when it is reduced.

● 小数のたし算・ひき算も整数と同じように，位ごとにたしたり，ひいたりします。

In the addition or subtraction of decimals, like integers, we add or subtract figures for each place.

● 小数と分数はコインの裏表のようなものです。

Decimals and fractions are things like both sides of a coin.

❾ 分数とわり算
Fractions and Division

＊分数の読み方と書き方については巻末の「解説」をご覧ください。

分数の名前

分数 …………………………………………[vulgar] fraction

$\dfrac{3}{5}$ ← 分子………numerator
　　　← 分母………denominator

$\dfrac{3}{5}$ ← 真分数………proper fraction

$\dfrac{11}{3}$ ← 仮分数………improper fraction

$3\dfrac{2}{7}$ ← 帯分数………mixed number

逆数 …………………………………………reciprocal
〜を逆数にする ……………change 〜 to its reciprocal

$\dfrac{3}{5} \longleftrightarrow \dfrac{5}{3}$　　　$\dfrac{12}{7} \longleftrightarrow \dfrac{7}{12}$

約分と通分

約分 ……canceling (cancelling 【英】) [down]／
　　　　cancellation／reduction

〜を約分する
　………cancel down 〜 to [its] lowest terms／
　　　　reduce 〜 to [its] lowest terms

$$\frac{8}{24} = \frac{4}{12} = \frac{2}{6} = \frac{1}{3}$$

$$\frac{12}{16} = \frac{9}{12} = \frac{6}{8} = \frac{3}{4}$$

通分 ……change of fractions with a common
　　　　denominator

〜を通分する
　………change 〜 to [equivalent] fractions with a
　　　　common denominator

公分母……………………………………common denominators

最小公分母
　………the least (lowest) common denominator (LCD)

$$\frac{2}{3} \longrightarrow \frac{4}{6}, \frac{6}{9}, \frac{8}{12}, \frac{10}{15}, \cdots\cdots$$

$$\frac{3}{4} \longrightarrow \frac{6}{8}, \frac{9}{12}, \frac{12}{16}, \frac{15}{20}, \cdots\cdots$$

分数の読み方

$\frac{1}{2}$ ················· one (a) half
$\frac{2}{3}$ ················· two thirds (two over three)
$\frac{4}{5}$ ················· four fifths (four over five)
$3\frac{2}{5}$ ················· three and two fifths
$\frac{9}{7}$ ················· nine sevenths (nine over seven)

●9の $\frac{2}{3}$ はいくらですか。
What is $\frac{2}{3}$ (two thirds) of 9 (nine)?

●3 $\frac{5}{12}$ を仮分数にしなさい。
Write $3\frac{5}{12}$ (three and five twelfths) as an improper fraction.

● $\frac{15}{4}$ を帯分数に直しなさい。
Change $\frac{15}{4}$ (fifteen fourths, fifteen over four) to a mixed number.

● $\frac{2}{3}$ と $\frac{3}{5}$ ではどちらが大きいですか。
Which of $\frac{2}{3}$ (two thirds) and $\frac{3}{5}$ (three fifths) is greater?

● $\frac{1}{2}$ は $\frac{3}{4}$ より小さいです。
$\frac{1}{2}$ (One half) is less than $\frac{3}{4}$ (three fourths).

● $\frac{5}{6}$ と $\frac{3}{4}$ を通分してください。
Change $\frac{5}{6}$ (five sixths) and $\frac{3}{4}$ (three fourths) to [equivalent]

fractions with a common denominator.

● $\frac{9}{12}$ を約分すると, $\frac{3}{4}$ になります。
$\frac{9}{12}$ (nine twelfths, nine over twelve) becomes $\frac{3}{4}$ (three fourths) when it is reduced.

●2$\frac{3}{4}$を小数にしなさい。
Write $2\frac{3}{4}$ (two and three fourths) as a decimal [number].

●4.25を分数に直しなさい。
Change 4.25 (four point two five) to a [vulgar] fraction.

●7の逆数は$\frac{1}{7}$です。
The reciprocal of 7 (seven) is $\frac{1}{7}$ (one seventh).

●3÷5は$\frac{3}{5}$です。
3 (Three) divided by 5 (five) is (equals) $\frac{3}{5}$ (three fifths).

●4mのヒモを5人でわけると，1人何mになりますか。
How many meters does one person get when a string of 4 m (four meters) is divided among 5 (five) people?

●分母と分子が同じときは，1と等しくなります。
When the denominator is the same as the numerator, the fraction is equal to 1 (one).

●分数も整数や小数と同じように数の1つです。
Like integers and decimals, fractions are a kind of numbers.

チャレンジ・テスト① CHALLENGE 練習問題

① 8（Eight）is two more than ☐ .
　［8は☐より2大きい。］

② 10（Ten）is three less than ☐ .
　［10は☐より3小さい。］

③ Is 23（twenty-three）less than or greater than 24（twenty-four）?
　［23は24より小さいですか，大きいですか。］

④ There are five color pencils in a row. What color is the 2nd (second) from the right?
　［色鉛筆が5本並んでいます。右から2番目は何色ですか。］

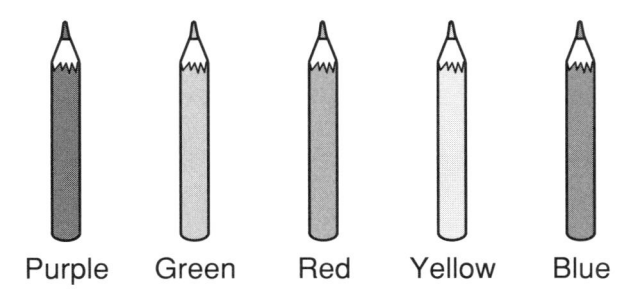

Purple　Green　Red　Yellow　Blue

⑤ What day of the week is it after 4 (four) days from Tuesday?
　［火曜日から4日先は何曜日ですか。］

⑥ What is the 2nd (second) digit of 2,345 (two thousand

three hundred and forty-five) from the bottom?
[2345の下から2桁目の数字はいくつですか。]

⑦ How many digits are in the number 879 (eight hundred [and] seventy-nine)?
[879は何桁の数ですか。]

⑧ An even number multiplied by any number is always ☐.
[偶数はどんな数をかけても ☐ になります。]

⑨ An odd number becomes an even number when multiplied by ☐, and becomes an odd number when multiplied by ☐.
[奇数は ☐ をかけると偶数になり, ☐ をかけると奇数になります。]

⑩ Write three common multiples of 4 (four) and 6 (six) starting with the smallest one.
[4と6の公倍数を小さい方から3つ書いてください。]

⑪ Write all the factors of 24 (twenty-four).
[24の約数をすべて書いてください。]

⑫ We can make a square by arranging 4 cm (four centimeter) by 5 cm (five centimeter) tiles. How many centimeters is the side of the smallest square?
[縦4cm, 横5cmのタイルを並べて正方形をつくります。いちばん小さい正方形の1辺は何cmでしょうか。]

⑬ We can divide 16 (sixteen) apples and 12 (twelve) oranges into equal groups with no remainders. How many people can share them? What is the largest number of people that can share them?
［りんご16個とみかん12個をどちらも余らないように同じ数ずつ配ろうと思います。何人に分けられますか。いちばん多いのは何人ですか。］

⑭ Any integer between A and B becomes 2,400 (two thousand four hundred) when rounded to the nearest hundred. What are A and B respectively?
［10の位を四捨五入して2400になるのはAからBまでの整数です。AとBはそれぞれいくつでしょうか。］

⑮ What number does 2,547 (two thousand, five hundred [and] forty-seven) become when rounded down to the nearest thousand?
［2547を100の位以下を切り捨てると，いくつになりますか。］

⑯ The number of visitors to the exhibition was 3,568 (three thousand five hundred sixty-eight). How many people does it become when rounded, rounded up and rounded down, respectively, to the nearest hundred?
［展覧会の入場者は3568人でした。10の位で四捨五入，切り上げ，切り捨てると，それぞれ何人になるでしょうか。］

⑰ List the numbers 7 (seven), 3 (three), 12 (twelve), 5 (five), and 9 (nine) in ascending order using the inequality sign.
［7，3，12，5，9を不等号を用いて小さい順に並べなさい。］

⑱ Which is larger, −14 (negative fourteen) or −23 (negative twenty-three)?
   ［−14と−23とではどちらが大きいですか。］

⑲ Indicate the numbers 5 (five), −6 (negative six), 0 (zero), 7 (seven), and −3 (negative three) on the number line.
   ［5, −6, 0, 7, −3を数直線を書いて示してください。］

⑳ The temperature was −3℃ (minus three degrees Celsius) yesterday and 4℃ (four degrees Celsius) today. What is the difference in temperature?
   ［昨日の気温は−3℃、今日は4℃です。温度の差はいくらでしょうか。］

㉑ The number accumulating four 2's (two's) and four 0.1's (zero point one's) is ☐ .
   ［2と0.1をそれぞれ4個集めた数は ☐ です。］

㉒ 4 mm (Four millimeters) is ☐ cm (centimeter).
   ［4mmは ☐ cmです。］

㉓ Read aloud 3.479.
   ［3.479を読んでください。］

㉔ Write $\frac{17}{3}$ (seventeen thirds) as a mixed number.
   ［$\frac{17}{3}$を帯分数で表わしましょう。］

㉕ Reduce $\frac{15}{55}$ (fifteen fifty-fifths) [to its lowest terms].
   ［$\frac{15}{55}$を約分しましょう。］

㉖ Write each number after A, B and C on the following number line both as a decimal and as a fraction.
　［つぎの数直線のA，B，Cにあたる数を小数と分数で表わしてください。］

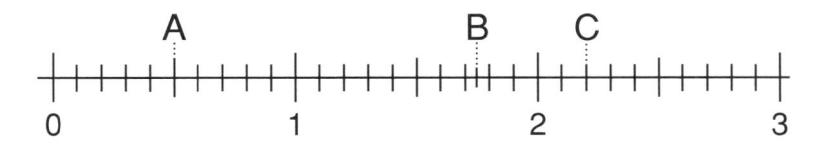

㉗ Change the fractions $\frac{11}{5}$ (eleven fifths), $\frac{5}{2}$ (five halves), and $\frac{7}{4}$ (seven quarters) to decimals, and list them in ascending order.
　［$\frac{11}{5}$，$\frac{5}{2}$，$\frac{7}{4}$ を小数に直し，小さい順に並べてください。］

Tea Room ❶

年齢・○歳代・年数

① − 人間や動物の年齢は数で表わします。
　　　　Tom is 12 (twelve) years old.
　　　　　──トムは12歳です。

② − 形容詞的用法の場合は複数のsを入れず，ハイフンを入れます。
　　　　a twelve-year-old boy（12歳の少年）

③ − ○歳○か月はandで結びます。
　　　　The baby is two years and six months old.
　　　　　──その赤ちゃんは2歳と6か月です。

④ − ○歳代の基本形はmy, your, her などとtwenties, thirties などとの組み合わせです。
　　　　Our best users are ladies in their twenties and thirties.
　　　　　──私たちのありがたい顧客は20代と30代の女性たちです。

⑤ − teenまたはteenagerは13〜19歳までを表わします。
　　　　Betty will be a teenager next month—13 at last.
　　　　　──ベティーは来月いよいよ13歳，ティーンエイジャーになります。

⑥ − 年齢層はage groupといい、下限と上限の年齢を数字で表わします。
　　　　We believe that this product is preferred in the age group ranging from 30 to 40.
　　　　　──この商品は30〜40歳までの年齢層に好まれると確信しています。

⑦ − 創立年数や建築年数など人間や動物以外の年数にも年齢と同じ表現が使えます。
　　　　Orange Corporation is an eight-year-old concern.
　　　　　──オレンジ社は設立8年の会社です。

第2章 数式と計算

Expressions and Operations

① たし算（整数・分数・小数）
Addition

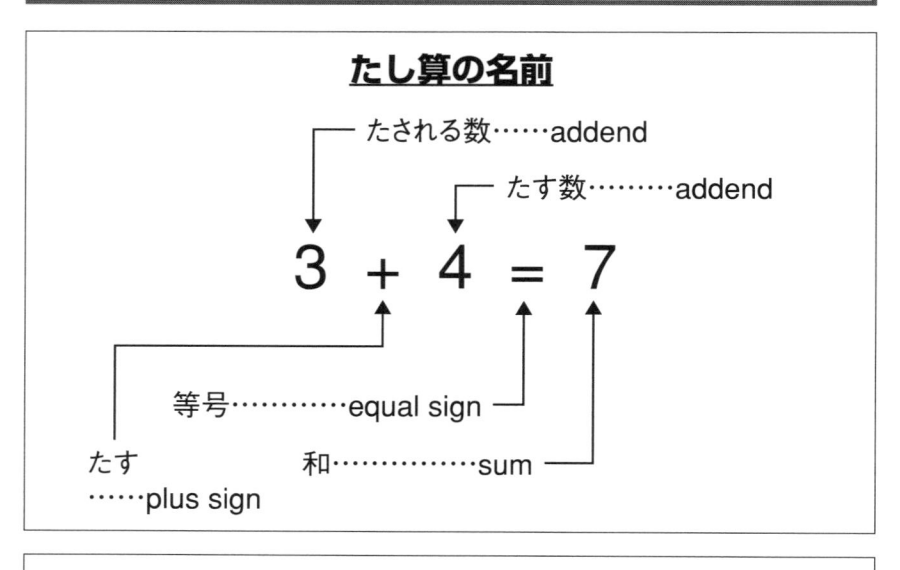

たし算の名前

たされる数……addend
たす数………addend

$$3 + 4 = 7$$

たす……plus sign
等号………equal sign
和…………sum

主要なことば

たし算………………addition
演算…………………operation
筆算…………………longhand calculation, longhand method
暗算…………………mental calculation, mental arithmetic

   * * *

くり上がり………………carrying
○に□をたす………add □ to ○
○と□をたす………add ○ and □
5に3をたす………add 3 (three) to 5 (five)
8と6をたす………add 8 (eight) and 6 (six)

たし算の読み方

○ + □ = △ ············ ○ plus □ equals △.

5 + 7 = 12
  ······5 (Five) plus 7 (seven) equals 12 (twelve).

3.7 + 4.6 = 8.3
  ······3.7 (Three and seven tenths) plus 4.6 (four and six tenths) equals 8.3 (eight and three tenths).

$\frac{4}{7} + \frac{2}{7} = \frac{6}{7}$
  ······$\frac{4}{7}$ (Four over seven) plus $\frac{2}{7}$ (two over seven) equals $\frac{6}{7}$ (six over seven).

●3に9をたしなさい。
Add 9 (nine) to 3 (three).

●24+34はいくつになりますか。
What is the result of 24+34 (twenty-four plus thirty-four)?

●7.9と2.3をあわせると，いくつですか。
What is the sum of 7.9 (seven point nine) and 2.3 (two point three)?

●$\frac{8}{9} + \frac{3}{8}$ を求めましょう。
Find the result of $\frac{8}{9} + \frac{3}{8}$ (eight ninths plus three eighths).

●120円のノートと205円の便箋と52円の消しゴムを買いました。全部でいくらでしょう。

第2章●数式と計算／Expressions and Operations

Someone bought a notebook for 120 (one hundred [and] twenty) yen, writing paper for 205 (two hundred [and] five) yen, and an eraser for 52 (fifty-two) yen. How much was it in total?

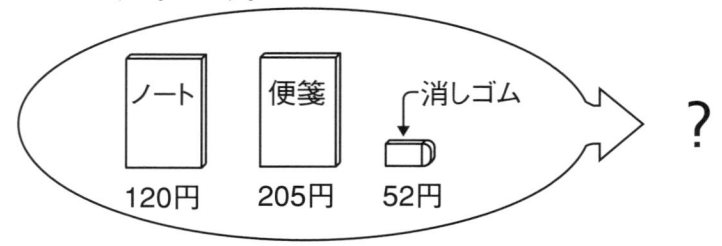

● お湯がポットに0.5ℓ，やかんに1.3ℓあります。あわせて何ℓですか。

There is 0.5 (zero point five) liter of hot water in the pot and 1.3 (one point three) liters in the kettle. How many liters is it in total?

● ジュースが $\frac{3}{4}$ ℓはいっているビンと $\frac{2}{5}$ ℓはいっているビンがあります。全部でいくらでしょう。

There are two bottles: one with $\frac{3}{4}$ (three fourths) liter of juice and another with $\frac{2}{5}$ (two fifths) liter. How many liters is it in total?

● たし算とひき算は，長さと重さのように異質なものの間ではできません。

We can neither add nor subtract two different types of things, such as length and weight.

● 分数のたし算とひき算は分母をそろえます。

To do additions or subtractions between fractions, change them to equivalent fractions with a common denominator.

❷ ひき算（整数・分数・小数）
Subtraction

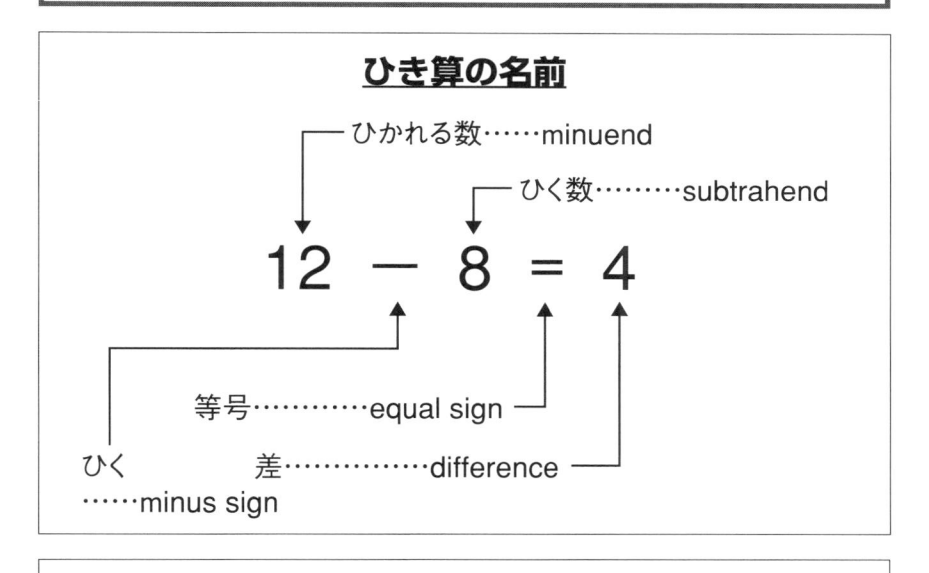

ひき算の名前

ひかれる数……minuend
ひく数………subtrahend

$$12 - 8 = 4$$

等号………equal sign
ひく……minus sign
差………difference

主要なことば

ひき算	subtraction
くり下がり	carrying down
○から□をひく	subtract □ from ○
○と□の差	the difference between ○ and □
残り	remainders
除去	subtraction
減少	reduction
差を求める（求差）	give differences
残りを求める（求残）	give remainders
おつり	change

第2章●数式と計算／Expressions and Operations

演算の関係

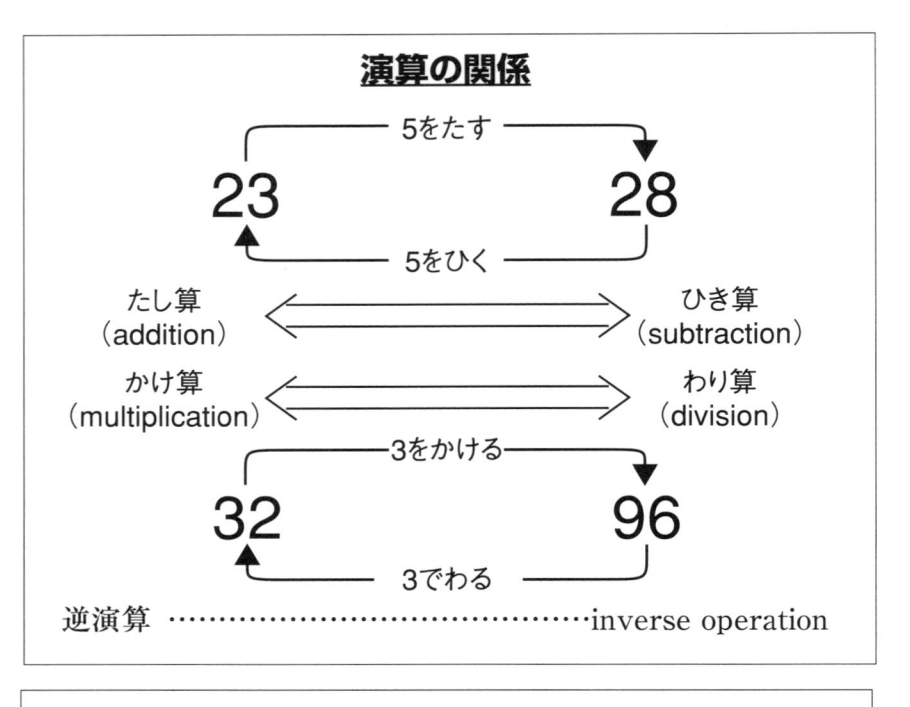

```
       ─── 5をたす ───▶
    23                28
       ◀── 5をひく ───
```

たし算（addition）　⟷　ひき算（subtraction）

かけ算（multiplication）　⟷　わり算（division）

```
       ─── 3をかける ──▶
    32                96
       ◀── 3でわる ───
```

逆演算 …………………………………… inverse operation

ひき算の読み方

○ − □ = △ …………………… ○ minus □ equals △.

63 − 24 = 39

　……63 (Sixty-three) minus 24 (twenty-four) equals 39 (thirty-nine).

8.4 − 2.5 = 5.9

　……8.4 (Eight and four tenths) minus 2.5 (two and five tenths) equals 5.9 (five and nine tenths).

$\frac{3}{4} - \frac{1}{4} = \frac{2}{4} = \frac{1}{2}$

　……$\frac{3}{4}$ (Three over four) minus $\frac{1}{4}$ (one over four) equals $\frac{2}{4}$ (two over four), which equals $\frac{1}{2}$ (one over two).

●84−63はいくつになりますか。

What is the result of 84−63 (eighty-four minus sixty-three)?

●20から15をひきなさい。

Subtract 15 (fifteen) from 20 (twenty).

●3.4と2.7の差はいくらでしょうか。

What is the difference between 3.4 (three point four) and 2.7 (two point seven)?

●$\frac{8}{9}$から$\frac{2}{9}$を除くと，残りはいくつですか。

What is the remainder when $\frac{2}{9}$ (two ninths) is subtracted from $\frac{8}{9}$ (eight ninths)?

●ここに46人います。28人が帰りました。残っているのは何人でしょうか。

There were 46 (forty-six) people here. 28 (Twenty-eight) of them have gone home. How many people are remaining?

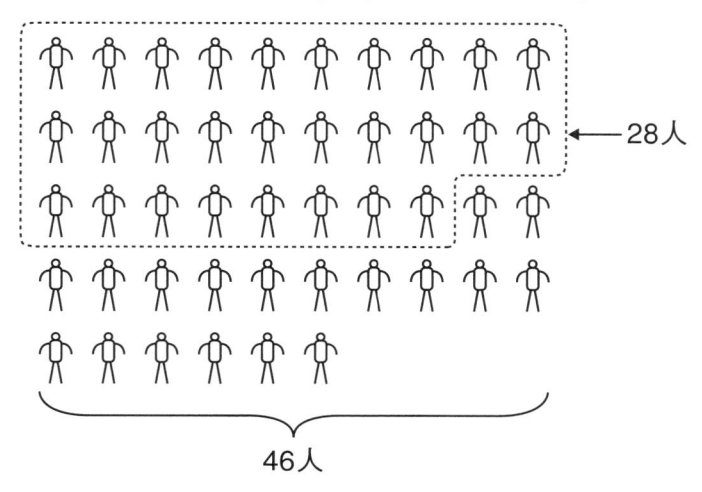

第2章●数式と計算／Expressions and Operations

●倉庫に5.5kgのお米と4.8kgの大豆があります。お米のほうがちょうど0.7kgだけ重いです。

There are 5.5 kg (five point five kilograms) of rice and 4.8 kg (four point eight kilograms) of soybean in the warehouse. The rice is just 0.7 kg (zero point seven kilogram) heavier.

●畑は $\frac{4}{5}$ a，田んぼは $\frac{7}{9}$ a です。どちらがどれだけ大きいでしょうか。

The field is $\frac{4}{5}$ a (four over five are) and the rice field is $\frac{7}{9}$ a (seven over nine are). Which is larger and by how much is it larger?

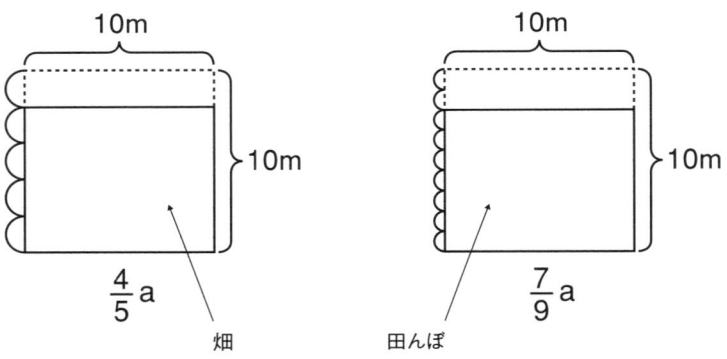

```
         10m                              10m
    ┌─────────┐                      ┌─────────┐
    │         │                      │         │
    │         │─10m                  │         │─10m
    │         │                      │         │
    └─────────┘                      └─────────┘
      4                                  7
      ─ a                                ─ a
      5                                  9
         畑                              田んぼ
```

●小数点をそろえ，位ごとに計算をします。

Align numbers by the decimal point, and then calculate for each place.

●ひき算には「差」を求める場合と「残り」を求める場合とがあります。

Subtractions give us "differences" in one case, and "remainders" in the other case.

❸ かけ算九九
Multiplication Tables

かけ算九九①

かけ算の九九表 ……………………multiplication table
　　　　　　＊　　　＊　　　＊
1 × 1 = 1　……1 (One) times 1 (one) equals 1 (one).
1 × 2 = 2　……1 (One) times 2 (two) equals 2 (two).
1 × 3 = 3　……1 (One) times 3 (three) equals 3 (three).
1 × 4 = 4　……1 (One) times 4 (four) equals 4 (four).
1 × 5 = 5　……1 (One) times 5 (five) equals 5 (five).
1 × 6 = 6　……1 (One) times 6 (six) equals 6 (six).
1 × 7 = 7　……1 (One) times 7 (seven) equals 7 (seven).
1 × 8 = 8　……1 (One) times 8 (eight) equals 8 (eight).
1 × 9 = 9　……1 (One) times 9 (nine) equals 9 (nine).
1 ×10 = 10　……1 (One) times 10 (ten) equals 10 (ten).
1 ×11 = 11　……1 (One) times 11 (eleven) equals 11 (eleven).
1 ×12 = 12　……1 (One) times 12 (twelve) equals 12 (twelve).

かけ算九九②

2 × 1 = 2　……2 (Two) times 1 (one) equals 2 (two).
2 ×12 = 24
　……2 (Two) times 12 (twelve) equals 24 (twenty-four).
3 × 3 = 9　……3 (Three) times 3 (three) equals 9 (nine).

$4 \times 1 = 4$ ……4 (Four) times 1 (one) equals 4 (four).
$9 \times 1 = 9$ ……9 (Nine) times 1 (one) equals 9 (nine).
$9 \times 9 = 81$
　……9 (Nine) times 9 (nine) equals 81 (eighty-one).
$9 \times 12 = 108$
　……9 (Nine) times 12 (twelve) equals 108 (one hundred and eight).

　　　　　＊　　　＊　　　＊

$11 \times 1 = 11$
　……11 (Eleven) times 1 (one) equals 11 (eleven).
$11 \times 12 = 132$
　……11 (Eleven) times 12 (twelve) equals 132 (one hundred [and] thirty-two).
$12 \times 11 = 132$
　……12 (Twelve) times 11 (eleven) equals 132 (one hundred [and] thirty-two).
$12 \times 12 = 144$
　……12 (Twelve) times 12 (twelve) equals 144 (one hundred [and] forty-four).

＊－欧米のかけ算九九表は日本と違い，12×12まであります。

●2の段と3の段をたすと，5の段になります。

Adding each numbers in the twice table and the three times table gives us each numbers in the five times table.

2の段	2	4	6	8	10	12	…
	+	+	+	+	+	+	
3の段	3	6	9	12	15	18	…
	‖	‖	‖	‖	‖	‖	
5の段	5	10	15	20	25	30	…

❹ かけ算 (整数・分数・小数)
Multiplication

かけ算の名前

かけられる数（被乗数）……multiplicand
かける数（乗数）……multiplier

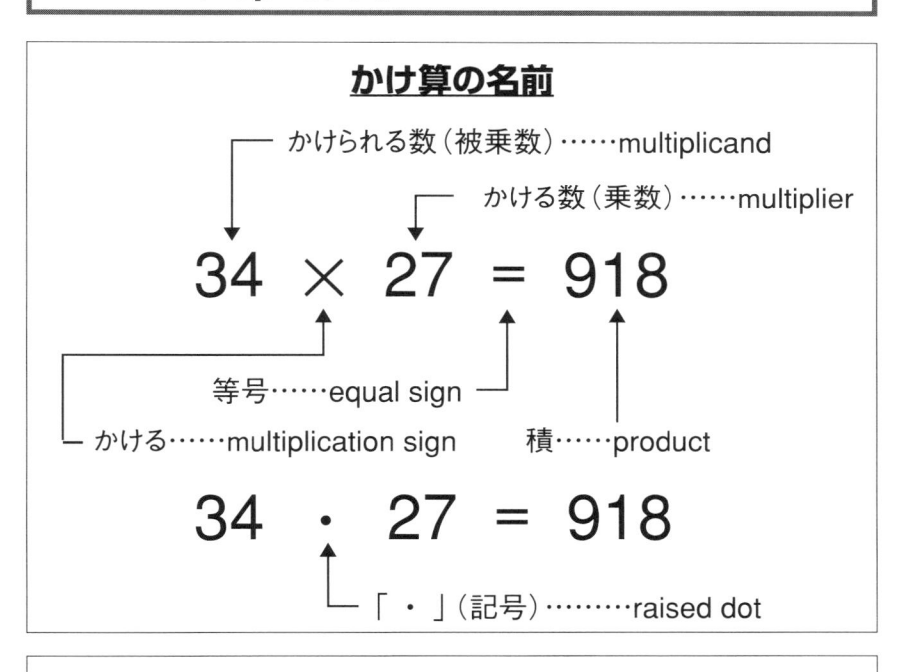

$$34 \times 27 = 918$$

等号……equal sign
かける……multiplication sign
積……product

$$34 \cdot 27 = 918$$

「・」（記号）……raised dot

主要なことば

かけ算　…………multiplication
○に□をかける　…multiply ○ by □
3に4をかける　…multiply 3 (three) by 4 (four)
4.3に5.6をかける
　……multiply 4.3 (four and three tenths) by 5.6 (five and six tenths)
$\frac{7}{9}$と$\frac{2}{5}$をかける
　……multiply $\frac{7}{9}$ (seven over nine) and $\frac{2}{5}$ (two over five)

4 の 3 倍 ……….3 (three) times 4 (four)／triple four
2.4 の 4 倍 ………4 (four) times 2.4 (two and four tenths)
$\frac{2}{6}$ の 5 倍 ……….5 (five) times $\frac{2}{6}$ (two over six)

かけ算の読み方

○ × □ = △ ……….○ times □ equals △.
26 × 6 = 156
　……26 (Twenty-six) times 6 (six) equals 156 (one hundred [and] fifty-six).
23 × 42 ……………23 (twenty-three) times 42 (forty-two)
0.4 × 2.6 ……………0.4 (point four) times 2.6 (two point six)
2.45 × 3.78
　……2.45 (two and forty-five hundredths) times 3.78 (three and seventy-eight hundredths)
$\frac{3}{7} \times \frac{5}{9}$
　……$\frac{3}{7}$ (three over seven) times $\frac{5}{9}$ (five over nine)
$2\frac{5}{8} \times 4\frac{3}{4}$
　……$2\frac{5}{8}$ (two and five eighths) times $4\frac{3}{4}$ (four and three fourths)

●24と32をかけると，いくつになりますか。
　What is the result when 24 (twenty-four) is multiplied by 32 (thirty-two)?

●52の7倍はいくつでしょう。

What is 7 (seven) times 52 (fifty-two)?

●16.2×32.4はいくらですか。

What is the result of 16.2×32.4 (sixteen point two times thirty-two point four)?

●$\frac{2}{5} \times 3\frac{4}{32}$ を求めなさい。

Find the result of $\frac{2}{5} \times 3\frac{4}{32}$ (two over five times three and four over thirty-two).

●1個234円のお菓子を1人に3個ずつ配りました。6人分の代金はいくらでしょう。

Three sweets costing 234 (two hundred [and] thirty-four) yen each have been distributed to each person. How much does it cost for six people?

●1m235円の布を4m買うと，いくら払えばいいでしょうか。

How much do I have to pay if I buy 4 m (four meters) of cloth, which costs 235 (two hundred [and] thirty-five) yen per meter?

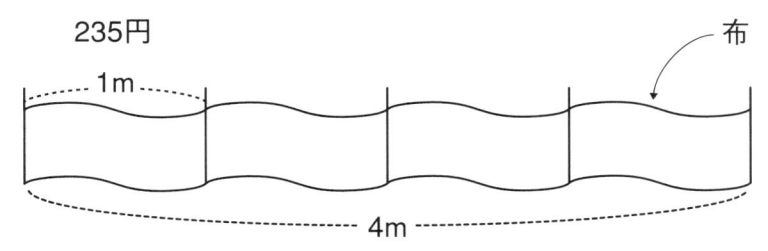

235円　　　　　　　　　　　　　　　　　　　布
1m
4m

●$2\frac{3}{4}$ mのロープを5本つくるのに，ロープは何m必要ですか。

How many meters of rope are needed to make five $2\frac{3}{4}$ m (two and three fourths meter) ropes?

第2章●数式と計算／Expressions and Operations

⑤ わり算（整数・分数・小数）
Division

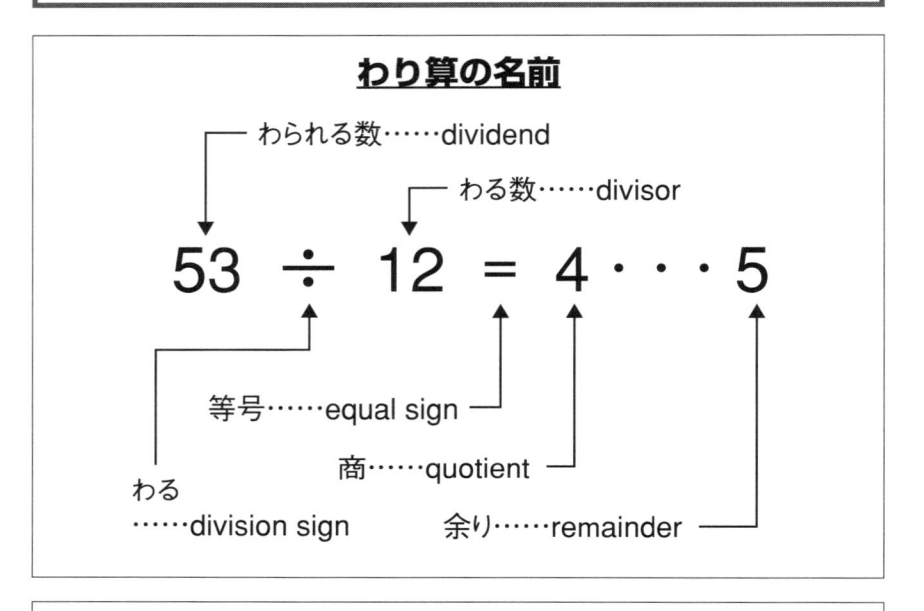

わり算の名前

$$53 \div 12 = 4 \cdots 5$$

- わられる数……dividend
- わる数……divisor
- 等号……equal sign
- 商……quotient
- わる……division sign
- 余り……remainder

主要なことば

わり算	division
○を□でわる	divide ○ by □
8を2でわる	divide 8 (eight) by 2 (two)
12は3でわりきれる	12 (Twelve) is divisible by 3 (three).
11は4でわりきれない	11 (Eleven) is not divisible by 4 (four).
6わる2は3です	Six divided by two is three.
計算の順序	the order of operations

わり算の読み方

○ ÷ □ = △ ················○ divided by □ equals △.

$12 \div 4 = 3$

　……12 (Twelve) divided by 4 (four) equals 3 (three).

$14 \div 4 = 3 \cdots 2$

　……14 (Fourteen) divided by 4 (four) equals 3 (three) with [a] remainder of 2 (two).

$3.57 \div 2.1 = 1.7$

　……3.57 (Three and fifty-seven hundredths) divided by 2.1 (two and one tenth) equals 1.7 (one and seven tenths).

$\frac{4}{9} \div \frac{2}{5} = \frac{10}{9} = 1\frac{1}{9}$

　……$\frac{4}{9}$ (Four ninths) divided by $\frac{2}{5}$ (two fifths) equals $\frac{10}{9}$ (ten ninths), which equals $1\frac{1}{9}$ (one and one ninth).

●20を5でわりなさい。

Divide 20 (twenty) by 5 (five).

●42わる7は6です。

42 (Forty-two) divided by 7 (seven) is 6 (six).

●398÷25を求めなさい。

Find the result of $398 \div 25$ (three hundred [and] ninety-eight divided by twenty-five).

● 5.6÷1.4はいくらですか。

What is the result of 5.6÷1.4 (five point six divided by one point four)?

● $\frac{4}{9} \div \frac{2}{3}$ は $\frac{2}{3}$ です。

$\frac{4}{9} \div \frac{2}{3}$ (Four over nine divided by two over three) is $\frac{2}{3}$ (two over three).

● 1本110円の缶ジュースがあります。890円で何本買えますか。

Each can of juice costs 110 (one hundred and ten) yen. How many cans can we buy with 890 (eight hundred and ninety) yen?

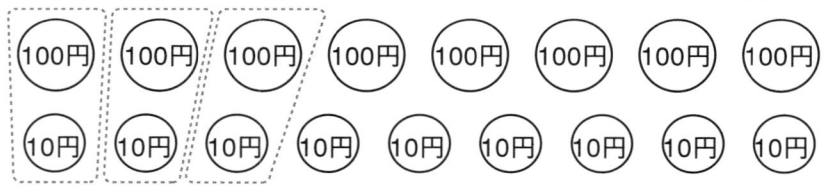

● 3.5gの砂糖を1袋に0.8gずつ入れます。何袋できて，何g余るでしょうか。

3.5 g (Three point five grams) of sugar will be divided into several bags each of which contains 0.8 g (zero point eight gram) of sugar. How many bags will be made, and how many grams of sugar will be left?

● $\frac{2}{3}$ dℓのペンキで $\frac{3}{4}$ m²の壁を塗ることができます。5 dℓでは何m²塗れるでしょうか。

$\frac{3}{4}$ (Three quarters) square meter of wall can be painted with $\frac{2}{3}$ dℓ (two thirds deciliter) of paint. How many square meters will be covered with 5 dℓ (five deciliters)?

❻ 四則混合算
Mixed Operations

四則と演算法則

四則 ……………………… four operations [in arithmetic]
 たし算 …………………… addition（＋）
 ひき算 …………………… subtraction（－）
 かけ算 …………………… multiplication（×）
 わり算 …………………… division（÷）
式 ………………………… expressions
括弧 ……………………… brackets, parentheses

    ＊ ＊ ＊

交換法則 ………………… commutative property
        the commutative law

$$● + ■ = ■ + ●$$
$$● × ■ = ■ × ●$$

結合法則 ………………… associative property
        the associative law

$$(● + ■) + ▲ = ● + (■ + ▲)$$
$$(● × ■) × ▲ = ● × (■ × ▲)$$

分配法則 ………………… distributive property
        the distributive law
括弧をはずす …………… expand (remove) brackets

$$(● + ▲) × ■ = ● × ■ + ▲ × ■$$

括弧でくくる ……………………bracket (parenthesize)

$$● × ■ - ▲ × ■ = (● - ▲) × ■$$

代入する ………………………substitute

$$x \overset{34}{+} 28 \ (= 62)$$

計算の読み方

$(120 + 84) × 7 = 1,428$
　……Opening bracket, one hundred [and] twenty plus eighty-four, closing bracket, times seven equals one thousand, four hundred [and] twenty-eight.

$240 × 3 - 32 × 5 = 560$
　……Two hundred [and] forty times three minus thirty-two times five equals five hundred [and] sixty.

$(45 - 21) ÷ 3 × 6 = 48$
　……Opening bracket, forty-five minus twenty-one, closing bracket, divided by three, times six equals forty-eight.

●太郎さんは1000円もって買い物に行きました。245円の牛乳と385円のお菓子を買いました。残りはいくらでしょうか。

　Taro went shopping with 1,000 (one thousand) yen. He bought milk for 245 (two hundred [and] forty-five) yen and sweets for 385 (three hundred [and] eighty-five) yen. How much money was left?

●シャツが定価の3500円より300円安くなっていました。3まい買うと，全部でいくらでしょう。

　300 (Three hundred) yen was deducted from the price of one 3,500 (three thousand [and] five hundred) -yen shirt. How much would it cost to buy three shirts at the reduced price?

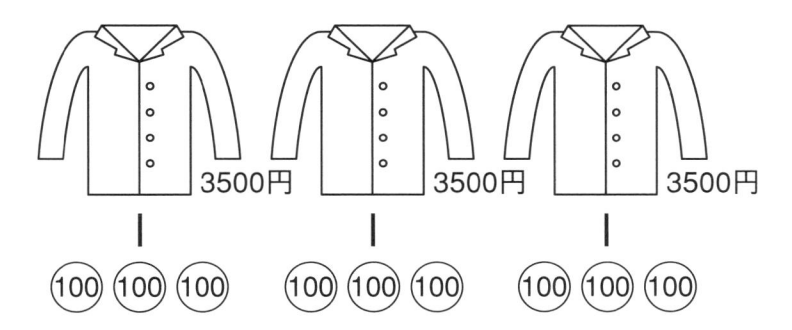

●1個250円のケーキ3個と，1個320円のプリン5個を買いました。1000円札を3枚だすと，おつりはいくらでしょうか。

　Someone bought three 250 (two hundred [and] fifty) -yen cakes and five 320 (three hundred [and] twenty) -yen caramel creams. How much is the change if he gives three 1,000 (one thousand) yen bills (notes)?

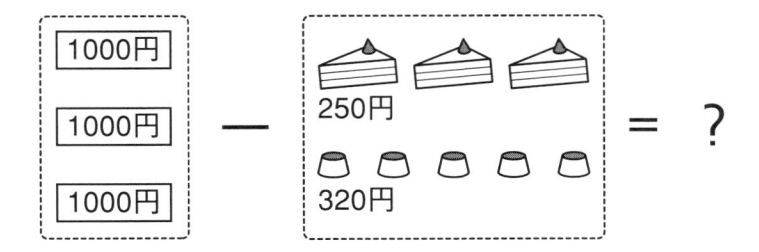

⑦ 文字と式

Variables and Expressions

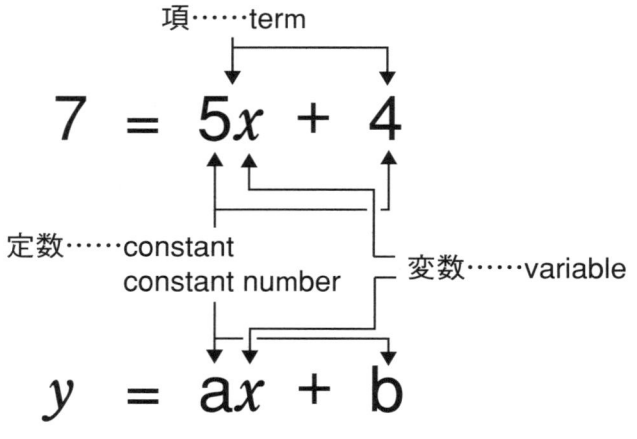

文字式の名前

項……term

$$7 = 5x + 4$$

定数……constant / constant number

変数……variable

$$y = ax + b$$

式……………………………………………expression

計算のきまり

$$3x+4y+5x+6y=(3+5)x+(4+6)y=8x+10y$$

同類項
(like (similar) terms)

同類項をまとめる
(combine like terms)

xに3を代入する……substitute 3 (three) for x

$$7x + 8 \ (= 29)$$

(with 3 substituted for x)

代入	substitution
置き換え（置換）	replacement
x を 5 で置き換える	replace x by (with) 5 (five)

計算の読み方

$4x$	four x
$x + 6 = 9$	x plus six equals nine.
$x - 8 = 23$	x minus eight equals twenty-three.
$6x \times 7 = 25$	Six x times seven equals twenty-five.
$8x \div 2 = 12$	Eight x divided by two equals twelve.
$x + y = 9$	x plus y equals nine.

●とりあえずわからない数を x にして方程式をたてます。

For the moment, let x be the unknown number and set up an equation.

●$3x+8$ の値を求めなさい。

Find the value of $3x+8$ (three x plus eight).

●$42-7x$ の値はいくつですか。

What is the value of $42-7x$ (forty-two minus seven x)?

●$11x$ に4をかけると，いくつでしょう。

What is the result of multiplying $11x$ (eleven x) by 4 (four)?

●$6x$ を2でわると，15になります。x はいくつでしょうか。

The result of dividing $6x$ (six x) by 2 (two) is 15 (fifteen). What is the value of x?

● 1個x円のケーキをy個買いました。式で表わしましょう。
We bought y pieces of x-yen cakes. Find an expression.

● 100円玉が同じ数だけはいっている箱が3つと，バラの100円玉が12個あります。全部をだして数えたら，27個ありました。1つの箱の中の100円玉の数をxとして式をたてなさい。
There were 3 (three) boxes, each of which contained the same number of 100-yen coins (one-hundred-yen coins), and twelve 100-yen coins (one-hundred-yen coins) which were loose. We counted 27 (twenty-seven) coins in total. Write an expression, letting x be the number of 100-yen coins (one-hundred-yen coins) in each box.

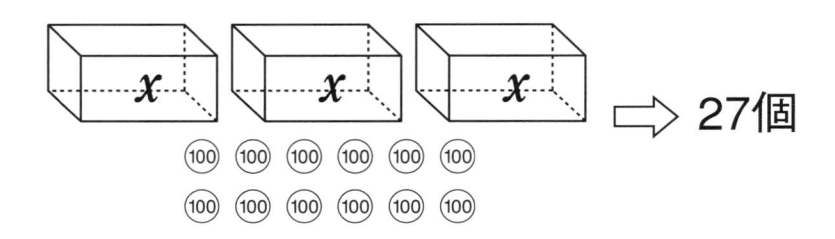

● パンが24個あります。お皿に同じ数だけ配ると，6皿になりました。お皿にわけたパンの数をxとして全部のパンの数についての方程式を書きましょう。
There were 24 (twenty-four) pieces of bread. They were distributed onto 6 (six) dishes with the same number of pieces on each dish. Write an equation for the total number of pieces, letting x be the number of pieces on each dish.

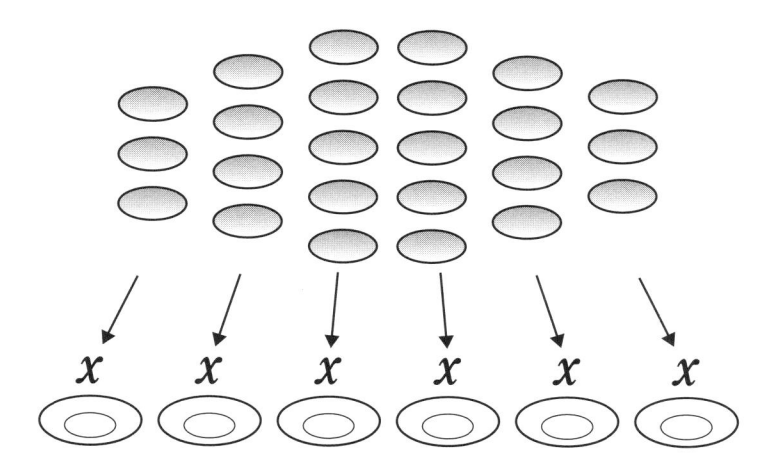

●チョコレートが6個はいっている箱の重さは1kgでした。箱の重さは460gです。チョコレート1つの重さをxとして，それを求める方程式を書きましょう。

The weight of a box containing 6 (six) chocolates was 1 kg (one kilogram). The weight of the box only was 460 g (four hundred [and] sixty grams). Let x be the weight of 1 (one) chocolate, then write an equation to find it.

チャレンジ・テスト② CHALLENGE 練習問題

① Add 7 (seven) to 3 (three). What is the result?
[3に7をたしなさい。それはいくつですか。]

② What is the sum of 21 (twenty-one) and 35 (thirty-five)?
[21と35の和はいくつですか。]

③ The number accumulating four 10's (ten's) and eight 1's (one's) is _____ .
[10を4つと1を8つ集めた数は_____です。]

④ There is a 3.6 m (three-point-six-meter) -tape and a 2.8 m (two-point-eight-meter) -tape. How many meters is it when the tapes are added together?
[3.6メートルと2.8メートルのテープがあります。あわせると，何メートルですか。]

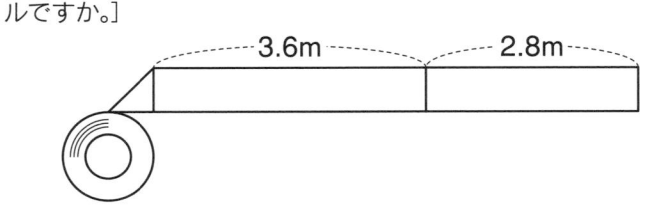

⑤ What is the result when 34 (thirty-four) is subtracted from 67 (sixty-seven)?
[67から34をひくと，いくつになりますか。]

⑥ There are 23 (twenty-three) red cards and 14 (fourteen) yellow ones. How many more red cards are there than yellow ones?

[赤いカードが23枚，黄色いのが14枚あります。赤いカードは黄色いのより何枚多いでしょうか。]

⑦ There were 46 (forty-six) children in the schoolyard and 28 (twenty-eight) among them have gone home. How many children remain? Write both an expression to show how many children remain and the answer.

[校庭に子どもが46人いましたが，28人が帰りました。何人残っているでしょうか。式と答えを書きましょう。]

⑧ There were 8 ℓ (eight liters) of water in a bucket, but 3 ℓ (three liters) spilled out. How much remain?

[バケツに8ℓの水がはいっていましたが，3ℓこぼしてしまいました。残りはいくらでしょうか。]

⑨ Mr. Suzuki weighs 48 kg (forty-eight kilograms), and Mr. Takahashi weighs 57 kg (fifty-seven kilograms). Which person is heavier, and by how much?

[鈴木君は体重が48kgで，高橋君は57kgです。どちらが，どれだけ重いでしょうか。]

⑩ Read aloud the following expressions:
2×7 4×8 15×11 23×32 42×108

[つぎの式を読んでください。]

⑪ There are 12 (twelve) months in one year. How many months are there in 4 (four) years?

[1年は12か月です。4年は何か月ですか。]

第2章●数式と計算／Expressions and Operations

⑫ How many oranges are there in all?
[みかんはぜんぶで何個あるでしょうか。]

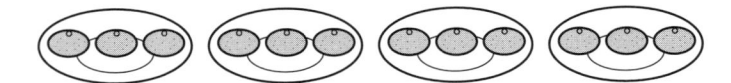

⑬ Mr. Yamada jogs for 30 (thirty) minutes every day. How many hours and minutes does he jog in a week?
[山田君は毎日30分ジョギングをします。1週間では何時間何分になりますか。]

⑭ 9 (Nine) teams, each of which consists of 7 (seven) people, ran a race. How many people ran in total?
[7人一組で9組が走りました。ぜんぶで何人が走りましたか。]

⑮ Divide 102 (one hundred and two) by 3 (three).
[102を3でわりなさい。]

⑯ Find the result of 9.2 (nine point two) divided by 2.3 (two point three).
[9.2÷2.3を求めなさい。]

⑰ There are 26 (twenty-six) peaches. If we divide them among 4 (four) people, how many does each person get, and how many are left over?
[桃が26個あります。4人で分けると，1人あたり何個で，余りはいくつになりますか。]

⑱ Make a story problem whose expression becomes 18÷

$3 = 6$.

[18÷3=6になる文章題をつくってみましょう。]

⑲ Find the values of each of the following expressions:
$82 - (38 - 45 ÷ 5)$ $125 × 4 + 36 ÷ 6$

[つぎの式の答えはいくらでしょうか。]

⑳ The price of an ¥850 (eight-hundred-and-fifty-yen) handkerchief is reduced by ¥65 (sixty-five yen). How much does it cost in total when buying 5 (five) of them?

[ハンカチが定価850円より65円安くなっていました。5枚買うと，全部でいくらでしょうか。]

㉑ Someone bought 3 (three) ¥250 (two hundred fifty yen) cakes and 4 (four) ¥350 (three hundred fifty yen) oranges. How much is the change when he gives 3 (three) ¥1,000 (one thousand yen) bills?

[250円のりんごを3個と350円のオレンジを4個買いました。1000円札を3枚だすと，おつりはいくらでしょうか。]

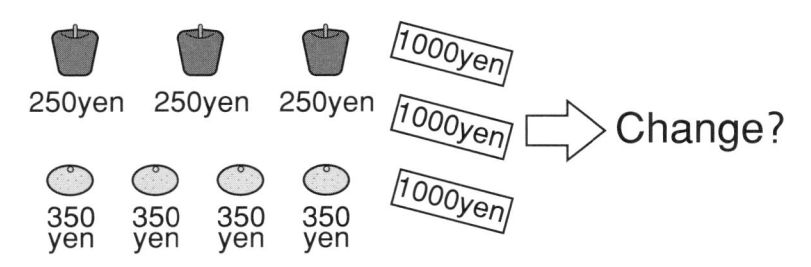

㉒ Find x in the equation: $7x + 8 = 50$

[7x+8=50のxを求めなさい。]

㉓ $8x$ (Eight x) multiplied by 6 (six) is 144 (one hundred forty-four). What is x?

[8xに6をかけると，144になります。x はいくつでしょうか。]

㉔ 3 (Three) bottles of $x\,\ell$ (x liter) juice divided among 5 (five) people resulted in 9 ℓ (nine liters) per person. Write an expression and find x.

[$x\,\ell$のジュース3本を5人で分けたら，1人あたり9ℓになりました。式で表わし，xを求めましょう。]

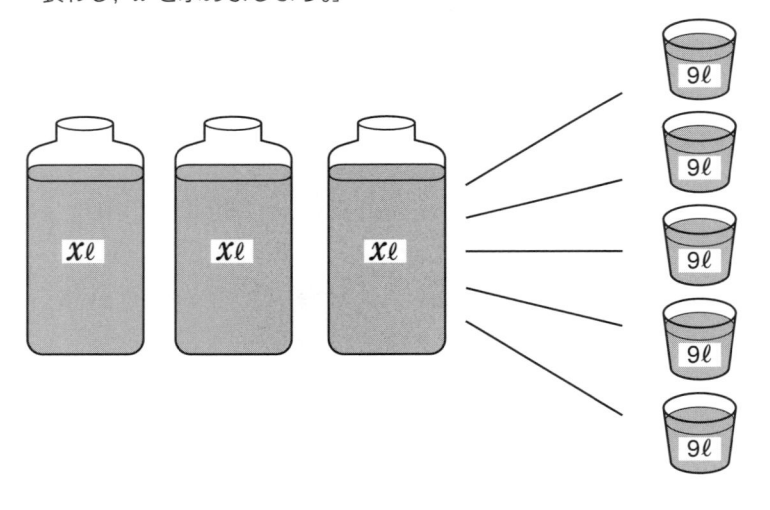

第 3 章
数量と単位

Numerical Quantities and Units

① 時間と時刻
Time and Duration

時間と時刻の名前

時間（一般）…………………time
時間（の長さ）
　……duration／period／interval／length of time
時刻……………………………time
時………………………………hour［記法 hr］／o'clock
分………………………………minute［記法 min］
秒………………………………second［記法 sec］
 1 時間…………………………an (one) hour
 1 分……………………………a (one) minute
 1 秒……………………………a (one) second
0.5時間…………………………half an hour (a half hour)
0.5分……………………………half a minute (a half minute)
0.5秒……………………………half a second (a half second)

時刻の読み方

○時□分△秒
　……○ [o'clock] □ minutes and △ seconds
午前 8 時16分 ………………eight sixteen a.m.
午後 2 時47分 ………………two forty-seven in the afternoon
 3 分24秒
　……3 (three) minutes 24 (twenty-four) seconds
 6 時間32分12秒

……6 (six) hours 32 (thirty-two) minutes 12 (twelve) seconds

4時8分すぎ
　　……8 (eight) minutes after (past) 4 (four)

5時13分まえ
　　……13 (thirteen) minutes of (to／until／till／before) 5 (five)

15時0分
　　……1500 (fifteen hundred) hours

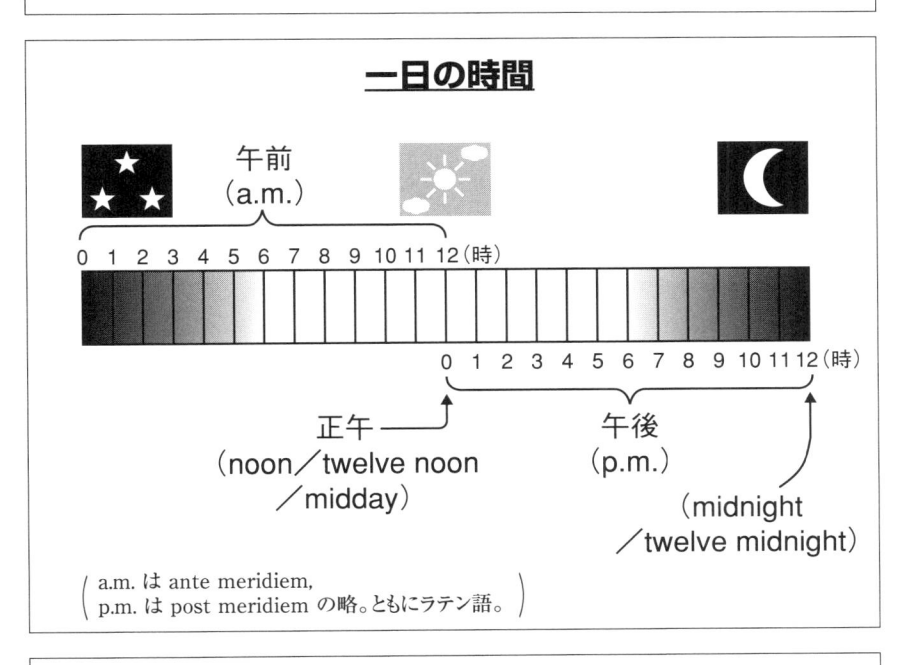

一日の時間

午前 (a.m.)

0 1 2 3 4 5 6 7 8 9 10 11 12 (時)

0 1 2 3 4 5 6 7 8 9 10 11 12 (時)

正午（noon／twelve noon／midday）

午後（p.m.）

(midnight／twelve midnight)

(a.m. は ante meridiem,
p.m. は post meridiem の略。ともにラテン語。)

時間の読み方

2時間半の道のり ………2.5 (two and a half) hours a way

3時から4時間後
　　……4 (four) hours after 3 (three) o'clock

10時から2時間半まえ
　……2.5 (two and [a] half) hours before 10 (ten) o'clock
2時から7時まで
　……from 2 (two) [o'clock] to 7 (seven)

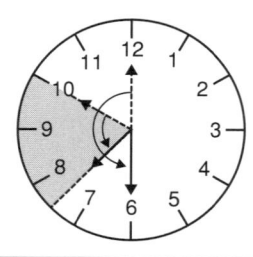

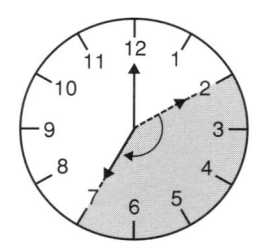

●午前9時8分に上町を出発した電車が，午後4時11分に下町に着きました。
　A train which left uptown at 9:08 (nine 〈ou〉 eight) a.m. arrived downtown at 4:11 (four eleven) p.m.

●ここから図書館まで25分くらいかかります。
　It takes about 25 (twenty-five) minutes from here to the library.

●あと15分で午後5時になります。
　It will be 5 (five) p.m. in 15 (fifteen) minutes.

●1分は60秒，1時間は60分，1日は24時間です。
　One minute is 60 (sixty) seconds, one hour is 60 (sixty) minutes, and one day is 24 (twenty-four) hours.

●午前と午後はそれぞれ12時間ずつあります。
　There are 12 (twelve) hours each in a.m. and in p.m.

❷ 長さ
Length

単位の名前

長さ ……………………length
道のり ……………………way
距離 ……………………distance
ミリメートル ……………millimeter【米】／millimetre【英】
　　　　　　　　　　　［記法 mm］
センチメートル …………centimeter【米】／centimetre【英】
　　　　　　　　　　　［記法 cm］
メートル …………………meter【米】／metre【英】
　　　　　　　　　　　［記法 m］
キロメートル ……………kilometer【米】／kilometre【英】
　　　　　　　　　　　［記法 km］

長さの読み方 【以下，米国表記】

○ m □ cm ………………○ meters □ centimeters
2 m 57 cm
　…………2 (two) meters 57 (fifty-seven) centimeters
4.8 m ……………………4.8 (four point eight) meters
4 cm 7 mm
　…………4 (four) centimeters 7 (seven) millimeters
3 km 76 m
　…………3 (three) kilometers 76 (seventy-six) meters

長さを比べる

237 cm = 2.37 m
　　　…………237 (Two hundred thirty-seven) centimeters
　　　　　　equals 2.37 (two point three seven) meters.
368 cm = 3 m 68 cm
　　　…………368 (Three hundred sixty-eight) centimeters
　　　　　　equals 3 (three) meters 68 (sixty-eight)
　　　　　　centimeters.
4 m ＜ 4.06 m
　　　…………4 (Four) meters is less than 4.06 (four point ⟨oh⟩
　　　　　　six) meters.
345 cm ＞ 322 cm
　　　…………345 (Three hundred [and] forty-five)
　　　　　　centimeters is greater than 322 (three hundred
　　　　　　[and] twenty-two) centimeters.

計算の読み方

3 m + 2 m 34 cm = 5 m 34 cm
　　　…………3 (Three) meters plus 2 (two) meters 34 (thirty-
　　　　　　four) centimeters equals 5 (five) meters 34
　　　　　　(thirty-four) centimeters.
7 m − 2 m 78 cm = 4 m 22 cm
　　　…………7 (Seven) meters minus 2 (two) meters 78
　　　　　　(seventy-eight) centimeters equals 4 (four)
　　　　　　meters 22 (twenty-two) centimeters.
5 cm × 6 = 30 cm
　　　…………5 (Five) centimeters times 6 (six) equals 30

(thirty) centimeters.

25 mm ÷ 5 = 5 mm
　………25 (Twenty-five) millimeters divided by 5 (five) equals 5 (five) millimeters.

●28cmから19cmをひくと，何cmになりますか。

How many centimeters result when 19 (nineteen) centimeters are subtracted from 28 (twenty-eight) centimeters?

●3km546mは何kmといえばいいですか。

How many kilometers can we say are in 3 (three) kilometers 546 (five hundred [and] forty-six) meters?

●家から郵便局まで870m，郵便局から駅まで790mあります。家から駅まで何mありますか。

It is 870 (eight hundred [and] seventy) meters from the house to the post office, and 790 (seven hundred [and] ninety) meters from the post office to the station. How many meters is it from the house to the station?

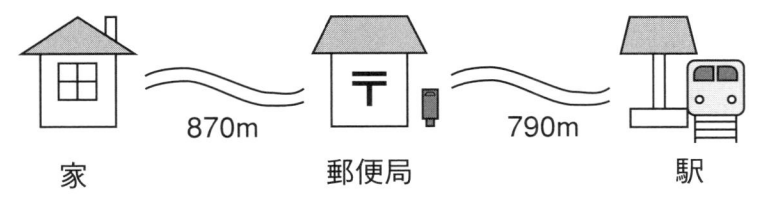

870m　　　790m
家　　　郵便局　　　駅

●4237mは何km何mといえますか。

How many kilometers and meters can we say are in 4,237 (four thousand two hundred thirty-seven) meters?

③ かさ（液量・容積）
Liquid Measurement

単位の名前

かさ（液量）	liquid measurement
かさ（容積）	measure of capacity
体積	volume
リットル	liter【米】／litre【英】
	［記法 l／L／ℓ］
ミリリットル	milliliter【米】／millilitre【英】
	［記法 mℓ］
デシリットル	deciliter【米】／decilitre【英】
	［記法 dℓ］
キロリットル	kiloliter【米】／kilolitre【英】
	［記法 kℓ］

かさの読み方　【以下，米国表記】

○ ℓ □ dℓ	○ liters □ deciliters
8 ℓ	8 (eight) liters
12 ℓ	12 (twelve) liters
2 ℓ 3 dℓ	2 (two) liters 3 (three) deciliters
34 dℓ	34 (thirty-four) deciliters
4.76 ℓ	4.76 (four point seven six) liters
100 dℓ	100 (one hundred) deciliters
345 mℓ	345 (three hundred [and] forty-five) milliliters

かさを比べる

30 $d\ell$ = 3 ℓ
　　　　…………30 (Thirty) deciliters equals 3 (three) liters.
1 ℓ = 10 $d\ell$ = 1000 $m\ell$
　　　　…………1 (One) liter equals 10 (ten) deciliters equals 1,000 (one thousand) milliliters.
24 ℓ < 247 $d\ell$
　　　　…………24 (Twenty-four) liters is less than 247 (two hundred [and] forty-seven) deciliters.
386 $d\ell$ > 38478 $m\ell$
　　　　…………386 (Three hundred [and] eighty-six) deciliters is greater than 38,478 (thirty-eight thousand, four hundred [and] seventy-eight) milliliters.

計算の読み方

215 $d\ell$ + 419 $d\ell$ = 634 $d\ell$
　　　　…………215 (Two hundred [and] fifteen) deciliters plus 419 (four hundred [and] nineteen) deciliters equals 634 (six hundred [and] thiry-four) deciliters.
67 ℓ − 12 ℓ = 55 ℓ
　　　　…………67 (Sixty-seven) liters minus 12 (twelve) liters equals 55 (fifty-five) liters.
42 $d\ell$ × 3 = 126 $d\ell$
　　　　…………42 (Forty-two) deciliters times 3 (three) equals 126 (one hundred [and] twenty-six) deciliters.
24 ℓ ÷ 4 = 6 ℓ

………24 (Twenty-four) liters divided by 4 (four) equals 6 (six) liters.

●5ℓ8dℓは何dℓですか。

How many deciliters is 5 (five) liters 8 (eight) deciliters?

●34500mℓは34.5ℓです。

34,500 (Thirty-four thousand five hundred) milliliters is 34.5 (thirty-four point five) liters.

●3ℓ5dℓ＋74dℓは何dℓですか。

How many deciliters is 3 (three) liters 5 (five) deciliters plus 74 (seventy-four) deciliters?

●67824mℓは67ℓ8dℓ24mℓです。

67,824 (Sixty-seven thousand eight hundred [and] twenty-four) milliliters is 67 (sixty-seven) liters 8 (eight) deciliters 24 (twenty-four) milliliters.

●456dℓは何ℓ何dℓですか。

How many liters and deciliters is 456 (four hundred [and] fifty-six) deciliters?

●2ℓのビーカーで3個，3dℓのビーカーで7個の水のかさはあわせて何ℓ何dℓですか。

How many liters and deciliters is the volume of water contained in three 2 (two)-liter beakers and seven 3 (three)-deciliter beakers in total?

●ジュースが3ℓ7dℓあります。2ℓ4dℓこぼしてしまいました。残りはいくらでしょうか。

There were 3 (three) liters 7 (seven) deciliters of juice. 2 (Two) liters 4 (four) deciliters have spilled. How much remains?

●500mℓのパックにはいっているミルクの3パック分は何ℓですか。

How many liters is the volume of three 500 (five hundred)-milliliter packs of milk?

●容積は体積と異なり，容器の内のりの縦・横・高さを使って求めます。

The capacity is different from the volume, and is found using the length, width and height of the inside measurement of a container.

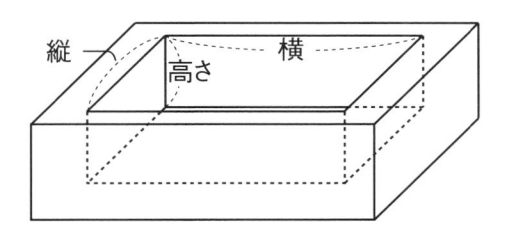

●かさは1ℓや1dℓなど単位を決めた容器のいくつ分かで比べます。

Capacities are measured by finding how many containers of a fixed unit, such as 1 (one) liter or 1 (one) deciliter, are equivalent.

●34ℓ5dℓと355dℓと35ℓをかさの大きい順に並べてください。

List in descending order 34 (thirty-four) liters 5 (five) deciliters, 355 (three hundred [and] fifty-five) deciliters and 35 (thirty-five) liters.

④ 重さ
Weight

単位の名前

重さ ……………………………………… weight
ミリグラム ……………………………… milligram ［記法 mg］
グラム …………………………………… gram ［記法 g］
キログラム ……………………………… kilogram ［記法 kg］

重さの読み方

2 kg 34 g ………2 (two) kilograms 34 (thirty-four) grams
5432 g
　　………5,432 (five thousand four hundred and thirty-two) grams

重さを比べる

3 kg 47 g ＝ 3047 g
　　………3 (Three) kilograms 47 (forty-seven) grams equals 3,047 (three thousand and forty-seven) grams.
3050 g ＝ 3.05 kg
　　………3,050 (Three thousand and fifty) grams equals 3.05 (three point 〈oh〉 five) kilo grams.
2695 g ＜ 3 kg
　　………2,695 (Two thousand six hundred ninety-five)

grams is less than 3 (three) kilograms.

2.6 kg＞2578 g
　　…………2.6 (Two point six) kilograms is greater than 2,578 (two thousand five hundred seventy-eight) grams.

計算の読み方

6 kg＋460 g ＝6460 g
　　…………6 (Six) kilograms plus 460 (four hundred sixty) grams equals 6460 (six thousand four hundred sixty) grams.

548 g －345 g ＝203 g
　　…………548 (Five hundred forty-eight) grams minus 345 (three hundred forty-five) grams equals 203 (two hundred and three) grams.

6 kg× 7 ＝42 kg
　　…………6 (Six) kilograms times 7 (seven) equals 42 (forty-two) kilograms.

56 g ÷ 7 ＝ 8 g
　　…………56 (Fifty-six) grams divided by 7 (seven) equals 8 (eight) grams.

●368ｇは何kgですか。
How many kilograms is 368 (three hundred sixty-eight) grams?

●2kg67ｇをｇで表わしてください。
Show 2 (two) kilograms 67 (sixty-seven) grams in grams.

●6789ｇは何kg何ｇといえますか。

How many kilograms and grams can we say are contained in 6,789 (six thousand seven hundred eighty-nine) grams?

●34kgに27kgをたしてください。

Add 27 (twenty-seven) killograms to 34 (thirty-four) kilograms.

●3040ｇ，3kg35ｇ，3kg80ｇを重い順に並べてください。

List in descending order by weight 3,040 (three thousand forty) grams, 3 (three) kilograms 35 (thirty-five) grams, and 3 (three) kilograms 80 (eighty) grams.

●私の体重は58kg，弟は42kgです。

I weigh 58 (fifty-eight) kilograms, and my brother weighs 42 (forty-two) kilograms.

●260ｇのお皿に砂糖を入れて測ったら，1kg 245ｇあります。砂糖の重さはいくらでしょう。

There is 1 (one) kilogram 245 (two hundred forty-five) grams when measuring sugar on a 260 (two hundred sixty)-gram dish. How much does the sugar weigh?

●2kgのみかん箱が3つと，3.3kgのりんご箱が4つあります。重さはあわせてどのくらいになりますか。

There are three 2 (two)-kilogram orange boxes and four 3.3 (three point three)-kilogram apple boxes. What is the total weight of all the boxes?

❺ 平均と密度と速さ
Average, Density and Speed

単位あたりの読み方

○あたり	per ○ （おもに書きことば）
	a ○ （おもに話しことば）
1メートルあたり	per meter
1グラムあたり	per gram
1日あたり	a (per) day
1時間あたり	an (per) hour
1平方メートルあたり	per square meter

主要なことば

単位	unit
メートル法	the metric system
平均（一般的な）	average
平均（算術的な）	[arithmetic] mean
概数	approximation
密度	density
人口密度	population density
濃度	concentration
速さ（速度）	speed／velocity
時速○キロメートル	○ kilometers per hour
分速○メートル	○ meters per minute
秒速○センチメートル	○centimeters per second

● 75ｇ，78ｇ，88ｇ，94ｇのトマトがあります。平均すると，1個何ｇでしょうか。

There is a 75 (seventy-five)-gram tomato, a 78 (seventy-eight)-gram tomato, an 88 (eighty-eight)-gram tomato, and a 94 (ninety-four)-gram tomato. What is the average weight of one tomato?

● 太郎さんが20歩歩いた距離は13ｍでした。歩幅はいくらでしたか。

The distance that Taro walked in 20 (twenty) steps was 13 (thirteen) meters. How long was each step?

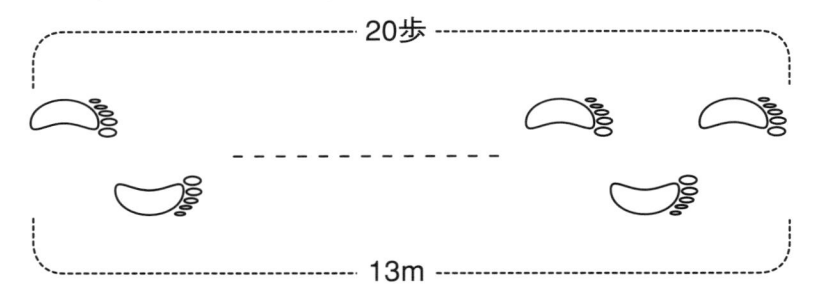

── 20歩 ──

── 13m ──

● 東京の面積は約2000ｋm²で，人口は約1000万人です。1km²あたりの人口密度は何人でしょう。

The area of Tokyo is approximately 2,000 (two thousand) square kilometers, and its population is about 10,000,000 (ten million). What is the population density per square kilometer?

● 電車は分速1.6kmで動いています。自動車は時速100kmで走っています。どちらがどれだけ速いでしょうか。

A train travels at 1.6 (one point six) kilometers per minute. A car travels at 100 (one hundred) kilometers per hour. Which is faster, and by how much?

●120km離れた町まで行くのに車で3時間かかりました。平均の速さはいくらですか。

It took 3 (three) hours to go to a town by car that was 120 (one hundred twenty) kilometers away. What was the average speed for the journey?

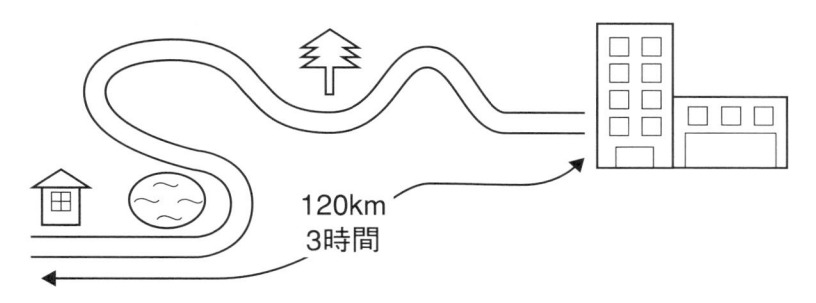

●音の秒速は340mです。稲妻が光ってから5秒後にカミナリの音が聞こえました。カミナリまでの距離は何kmですか。

The speed of sound is 340 (three hundred forty) meters per second. We heard the sound of thunder 5 (five) seconds after the flash of lightning. How many kilometers is the distance from here to the lightning?

チャレンジ・テスト③ / CHALLENGE 練習問題

① How many minutes and seconds is 124 (one hundred twenty-four) seconds?
[124秒は何分何秒ですか。]

② What time is it 50 (fifty) minutes after 8:15 (eight fifteen) a.m.?
[午前8時15分から50分後は何時何分ですか。]

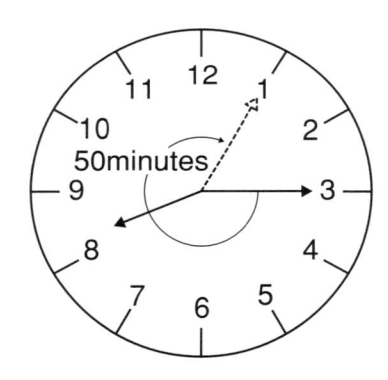

③ Takashi played from 2:30 p.m. to 4:40. For how many hours and minutes did he play?
[たかし君は午後2時半から4時40分まで遊びました。遊んだ時間は何時間何分ですか。]

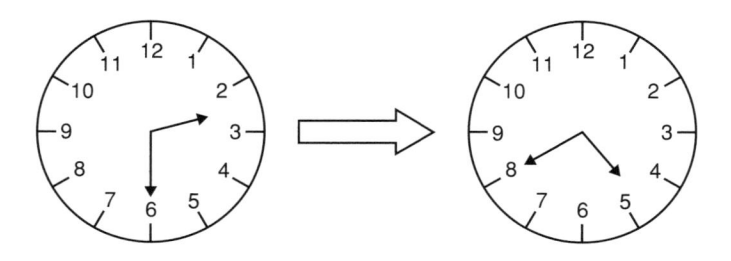

④ It takes 25 (twenty-five) minutes from home to the station. At what time will we arrive at the station if we leave home at 8:05 (eight ⟨ou⟩ five).
［家から駅まで25分かかります。8時5分に家をでると，駅に着くのは何時ですか。］

⑤ How many centimeters do we get when we add 21 cm (twenty-one centimeters) to 46 cm (forty-six centimeters) ?
［21cmと46cmをあわせると，何cmになりますか。］

⑥ Which is longer, and how much longer 4 m (four meters) or 403 cm (four hundred and three centimeters)?
［4mと403cmでは，どちらがどれだけ長いですか。］

⑦ How far is it from Mr. Yamada's house to the station?
［山田君の家から駅までの距離はどれくらいですか。］

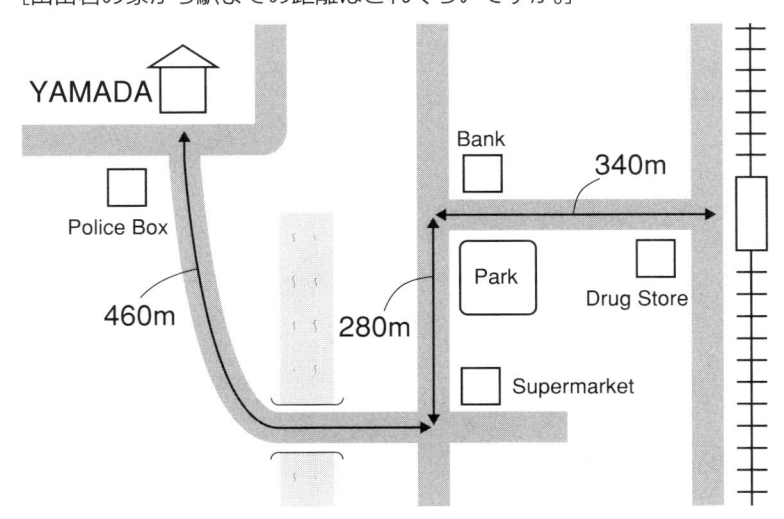

第3章●数量と単位／Numerical Quantities and Units

⑧ How many meters and centimeters do we get when we tie together 3 (three) strings 75 cm (seventy-five centimeters) long and 1 (one) string 2 m (two meters) long?
［75cmのヒモ3本と2mのヒモを結ぶと，何m何cmになりますか。］

⑨ How many deciliters are in 6 ℓ 8 $d\ell$ (six liters eight deciliters)?
［6 ℓ 8 $d\ell$ は何 $d\ell$ ですか。］

⑩ How many deciliters of water do we get when we add seven 1 (one)-liter cups of water to three 1 (one)-deciliter cups of water? How many liters are there?
［1 ℓ ますで7杯，1 $d\ell$ ますで3杯の水の量は何 $d\ell$ ですか。また，それは何 ℓ といえますか。］

⑪ How many liters are contained in 7 (seven) cartons of milk, each of which is a 500 $m\ell$ (five hundred milliliter) carton?
［500 $m\ell$ パックにはいっている牛乳7パック分は何 ℓ ですか。］

⑫ A container including 2 ℓ 8 $d\ell$ (two liters eight deciliters) of juice was knocked over and only 9 $d\ell$ (nine deciliters) remained. How many liters of juice spilled out?
［ジュースが2 ℓ と8$d\ell$はいっている容器を倒してしまい，9$d\ell$残りました。こぼしたのは何 ℓ ですか。］

⑬ 48 ℓ (Forty-eight liters) of coffee is to be divided evenly into 4 (four) containers. How many liters will each container get?

[48ℓのコーヒーを4個の容器にわけると，1個あたり何ℓになりますか。]

⑭ How many grams is 3 kg 52 g (three kilograms fifty-two grams)?

[3kg52gは何gですか。]

⑮ An elder brother weighs 53 kg (fifty-three kilograms), and his younger brother weighs 46 kg (forty-six kilograms). How much heavier is the elder brother?

[兄の体重は53kg，弟は46kgです。兄のほうがどれくらい重いですか。]

⑯ List 4056 g (four thousand fifty-six grams), 4 kg 34 g (four kilograms thirty-four grams) and 4 kg 87 g (four kilograms eighty-seven grams) in ascending order.

[4056g，4kg34g，4kg87gを軽い順に並べてください。]

⑰ The result of weighing the salt on a plate weighing 230 g (two hundred thirty grams) is 1 kg 32 g (one kilogram thirty-two grams). How much does the salt weigh?

[230gのお皿に塩をのせて測ったら，1kg32gありました。塩の重さはいくらでしょうか。]

⑱ There are five apples whose weights are 78 g (seventy-eight grams), 82 g (eighty-two grams), 85 g (eighty-five grams), 91 g (ninety-one grams), and 94 g (ninety-four grams). What is the average weight per apple?

[78g，82g，85g，91g，94gの5つのりんごがあります。平均の重さは，1個あたりいくらになりますか。]

⑲ 144 (One hundred and forty-four) flowers are planted with equal space between them in a flower bed of 12 m² (twelve square meters). How many flowers are there per square meter?

[12m²の花壇に花が同じ間隔で144本植えられています。1m²あたり何本ですか。]

⑳ Which is cheaper, and how much cheaper is it per meter, a cloth priced at ¥360 (three hundred sixty yen) for 3 m (three meters) or a cloth priced at ¥750 (seven hundred fifty yen) for 5 m (five meters)?

[3mで360円の布と5mで750円の布とでは，1mあたりの値段はどちらがいくら安いですか。]

㉑ A car ran 255 km (two hundred fifty-five kilometers) for 3 (three) hours. How many kilometers per hour was its speed?

[自動車が3時間に255km走りました。時速は何kmでしたか。]

第4章 平面図形と空間図形

Plane Figures and Solid Figures

① 点と線 （垂直・平行）
Points and Lines

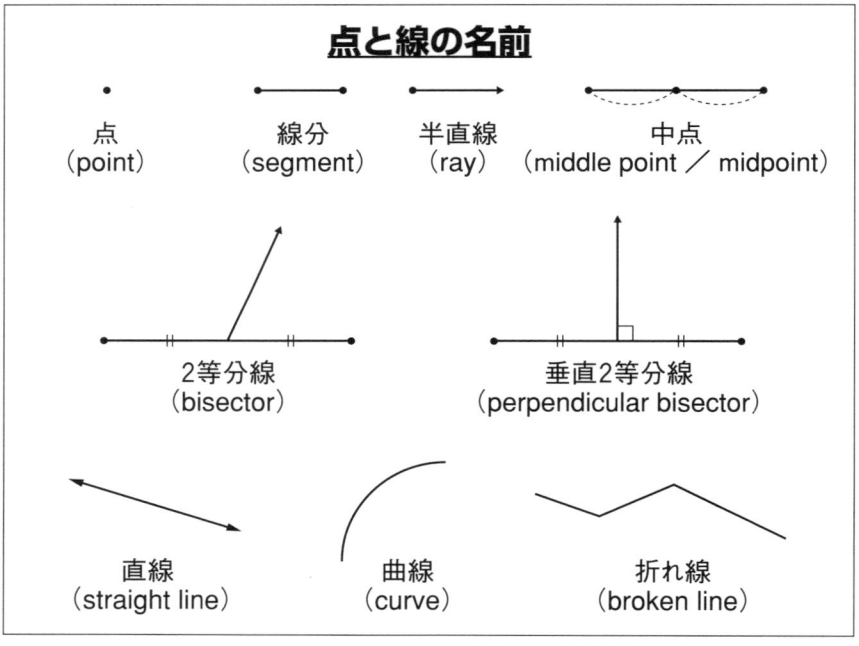

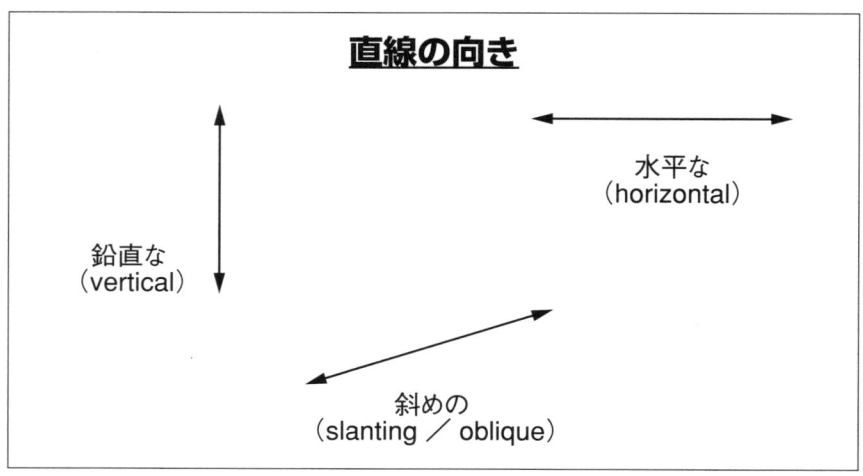

点と線の名前

- 点 (point)
- 線分 (segment)
- 半直線 (ray)
- 中点 (middle point／midpoint)
- 2等分線 (bisector)
- 垂直2等分線 (perpendicular bisector)
- 直線 (straight line)
- 曲線 (curve)
- 折れ線 (broken line)

直線の向き

- 鉛直な (vertical)
- 水平な (horizontal)
- 斜めの (slanting／oblique)

直線の関係

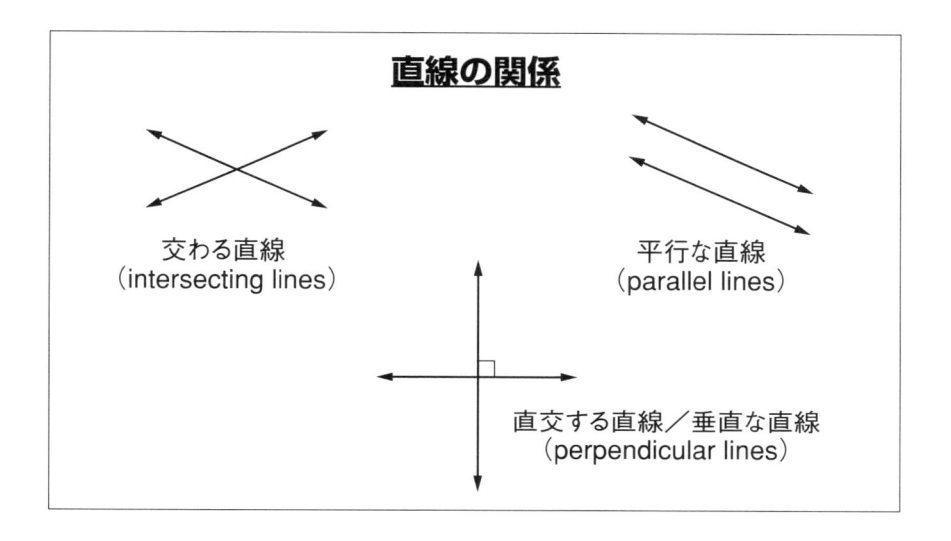

交わる直線
(intersecting lines)

平行な直線
(parallel lines)

直交する直線／垂直な直線
(perpendicular lines)

●線と線が交わる点を交点といいます。

A point where two lines intersect is called an intersection.

●直線Aは直線Bと直交しています。

Straight line A is perpendicular to straight line B.

●線分の中点を通る垂直な直線を垂直2等分線といいます。

The straight line which both contains the midpoint of and is perpendicular to a segment is called the perpendicular bisector of the segment.

●直角に交わる直線を垂直であるといいます。

Two straight lines are called perpendicular when they cross each other at a right angle.

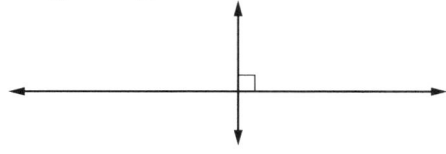

●1本の直線に平行な2本の直線は平行です。

Two straight lines that are parallel to another straight line are parallel.

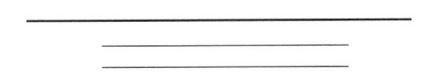

●平行な直線は幅がどこも等しく，どこまで延ばしても絶対に交わりません。

Parallel straight lines are the same distance apart everywhere and never cross each other however long they extend.

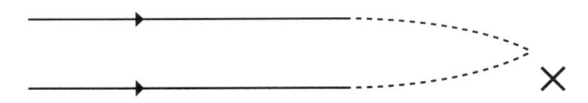

●平行な直線はほかの直線と等しい角度で交わります。

Parallel straight lines cross another straight line at the same angle.

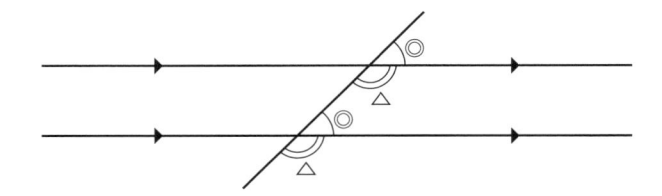

❷ 角と角度

Angles and their Measure

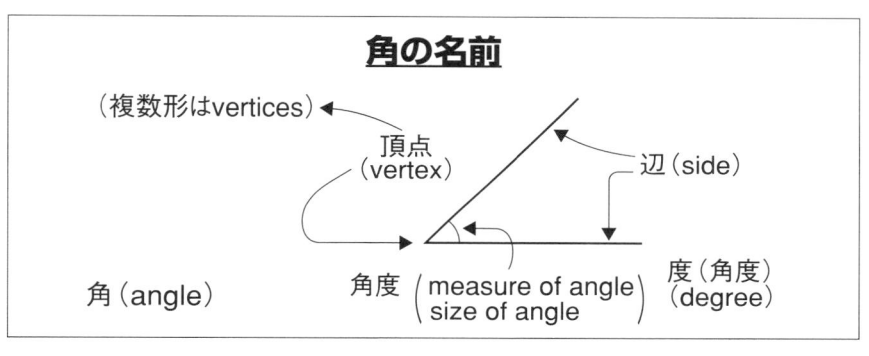

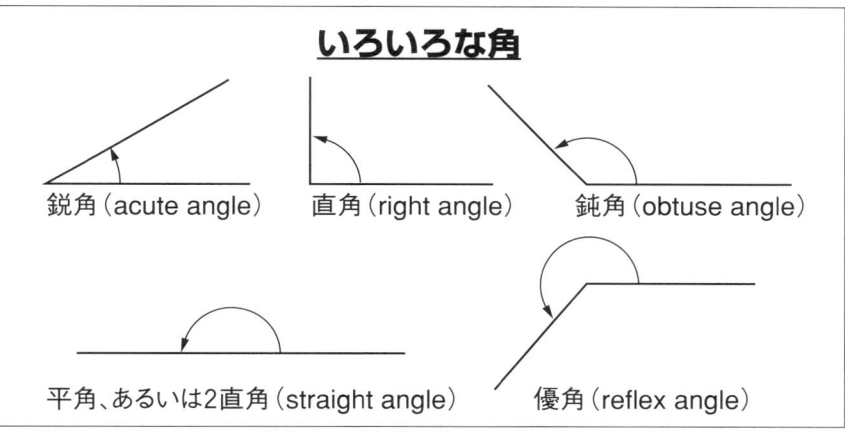

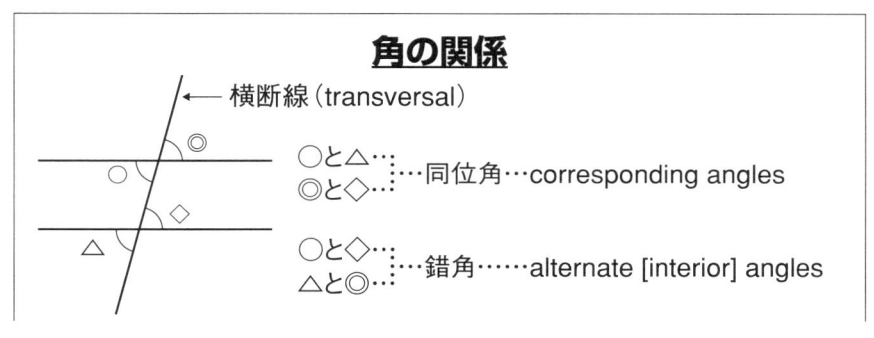

角の名前

（複数形はvertices）
頂点（vertex）
辺（side）
角（angle）
角度（measure of angle / size of angle）
度（角度）（degree）

いろいろな角

鋭角（acute angle）　直角（right angle）　鈍角（obtuse angle）

平角、あるいは2直角（straight angle）　優角（reflex angle）

角の関係

横断線（transversal）

○と△…
◎と◇… ……同位角…corresponding angles

○と◇…
△と◎… ……錯角……alternate [interior] angles

第4章●平面図形と空間図形／Plane Figures and Solid Figures

○と△：隣接角
(adjacent angles)

◎と□：対頂角
(vertical angles)

xとxx：余角
(complementary angles)
(complements)

＊と＊＊：補角
(supplementary angles)
(supplements)

角と直線

角の2等分線
(the bisector of an angle)

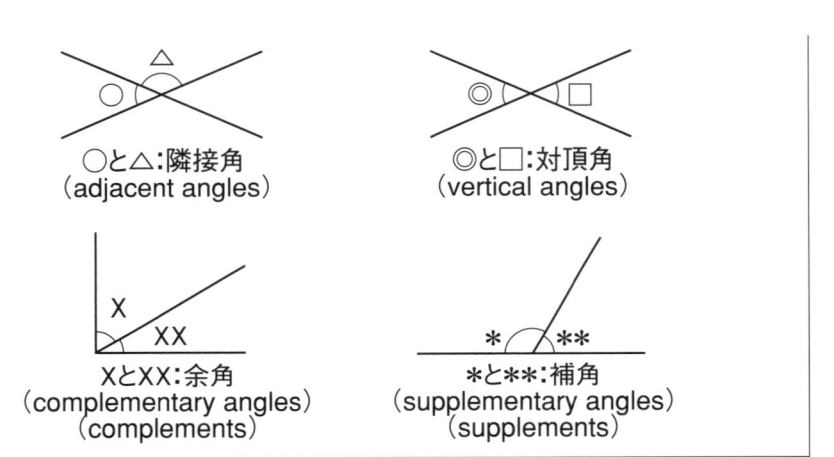

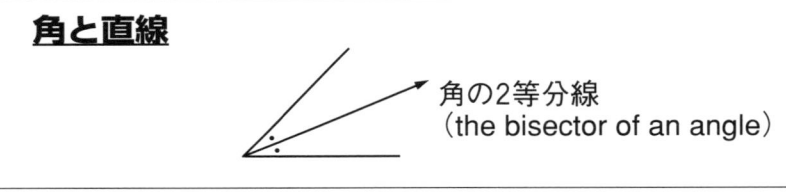

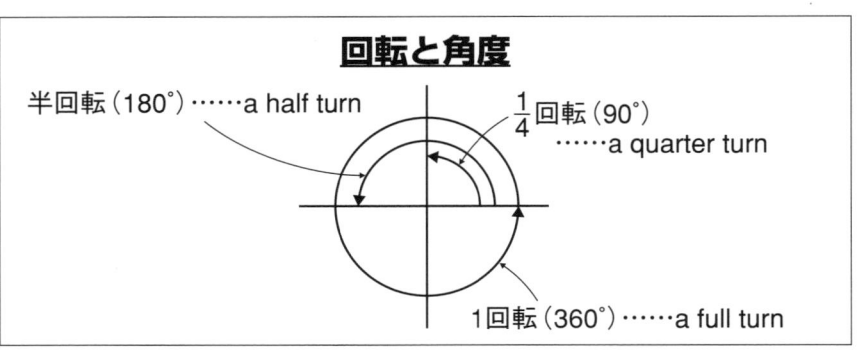

回転と角度

半回転（180°）……a half turn

$\frac{1}{4}$回転（90°）……a quarter turn

1回転（360°）……a full turn

角度の読み方

  0°　（度）…………zero degree
 45°　（度）…………forty-five degrees
 90°　（度）…………ninety degrees
180°　（度）…………one hundred eighty degrees

●角の大きさは角度で測ります。

The size of an angle is measured in degrees.

●分度器を使わずにつぎの角度を求めましょう。

Let's find the following angle without using a protractor.

●三角定規を下図のように組合せました。∠Aと∠Bはそれぞれ何度でしょうか。

Triangular rulers were combined together as in the following figure. How many degrees are angles A and B, respectively?

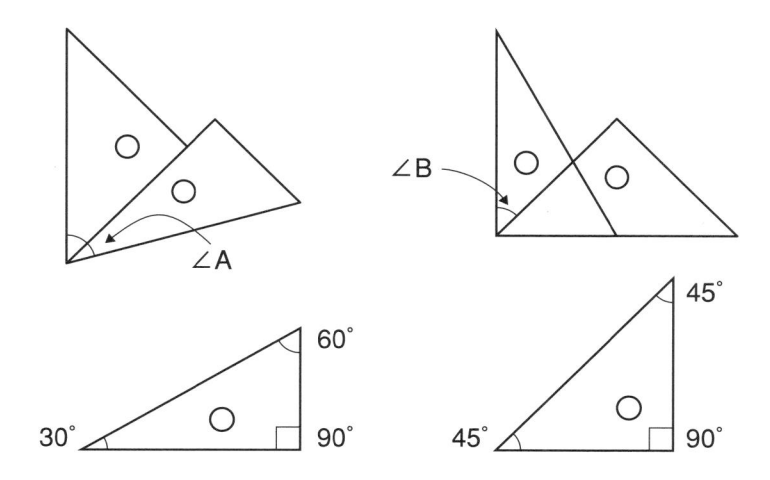

第4章●平面図形と空間図形／Plane Figures and Solid Figures

③ 3角形
Triangles

3角形の名前

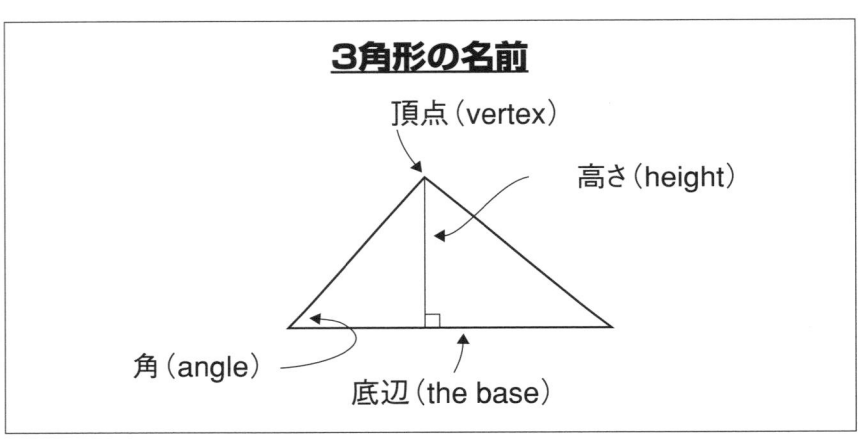

- 頂点（vertex）
- 高さ（height）
- 角（angle）
- 底辺（the base）

いろいろな3角形

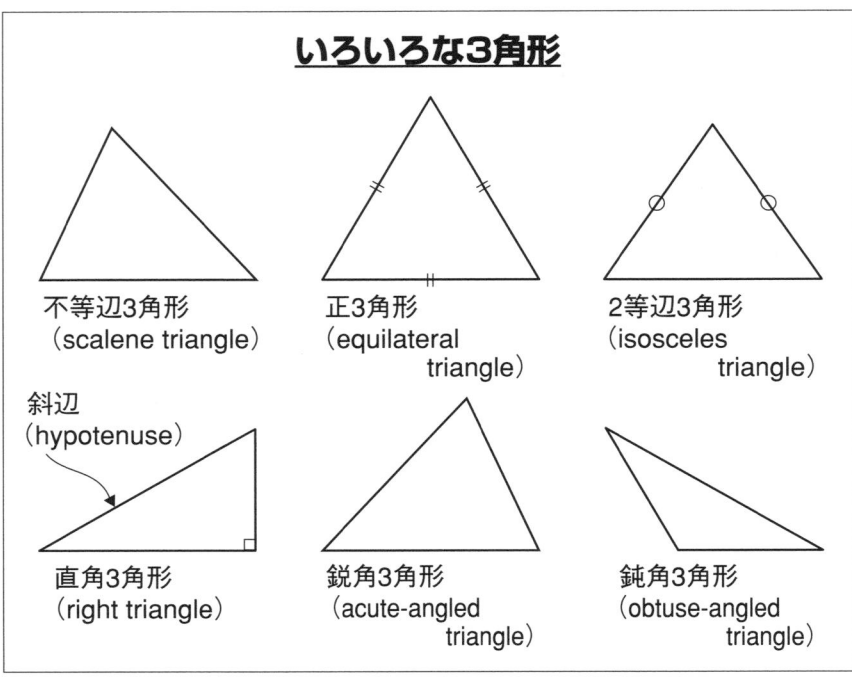

- 不等辺3角形（scalene triangle）
- 正3角形（equilateral triangle）
- 2等辺3角形（isosceles triangle）
- 斜辺（hypotenuse）
- 直角3角形（right triangle）
- 鋭角3角形（acute-angled triangle）
- 鈍角3角形（obtuse-angled triangle）

3角形と直線

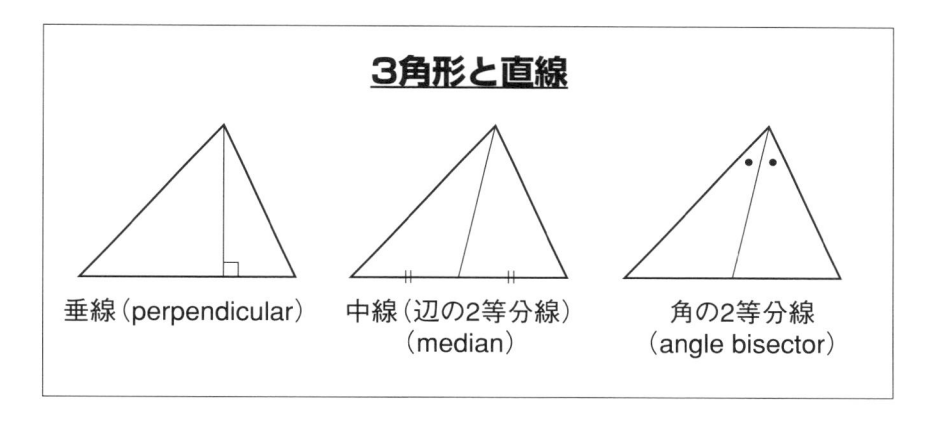

垂線（perpendicular）　　中線（辺の2等分線）　　角の2等分線
　　　　　　　　　　　　（median）　　　　　　（angle bisector）

●2つの辺が等しい3角形を2等辺3角形といいます。

A triangle whose two sides are equal is called an isosceles triangle.

●3つの辺の長さが等しい3角形を正3角形といいます。

A triangle whose three sides are equal in length is called an equilateral triangle.

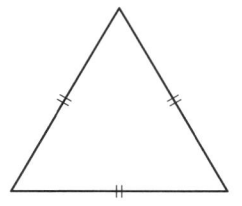

●2等辺3角形では2つの辺の長さが同じです。

In an isosceles triangle two sides have the same length.

● 正3角形では3つの角の大きさが同じです。

In an equilateral triangle three angles are the same size.

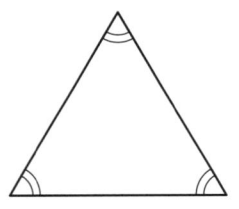

● 3角形の内角の和は180°です。

The sum of the interior angles of any triangle is 180 (one hundred eighty) degrees.

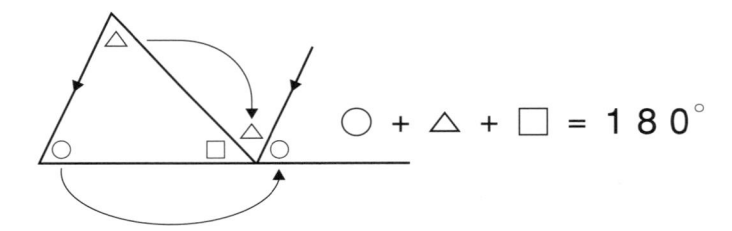

○ + △ + □ = 180°

● 3角形の外角の和は360°です。

The sum of the exterior angles of any triangle is 360 (three hundred sixty) degrees.

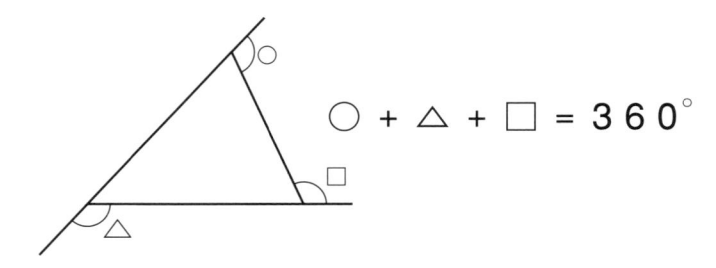

○ + △ + □ = 360°

④ 4角形（正方形・長方形・台形・平行4辺形・ひし形）
Quadrilaterals

いろいろな4角形

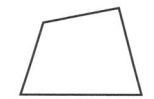

- 4角形（quadrilateral）
- 不等辺4角形（trapezoid【英】／trapezium【米】） 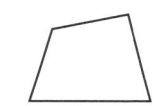
- 正方形（square）
- 長方形（rectangle）  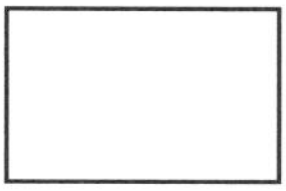
- 平行4辺形（parallelogram） 
- ひし形（rhombus） 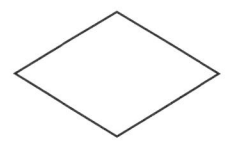
- 台形（trapezium【英】／trapezoid【米】） 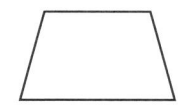
- たこ形（kite） 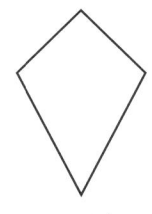
- へこんだたこ形（arrowhead） 
- 等脚台形（isosceles trapezium【英】／isosceles trapezoid【米】）

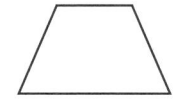

4角形の名前

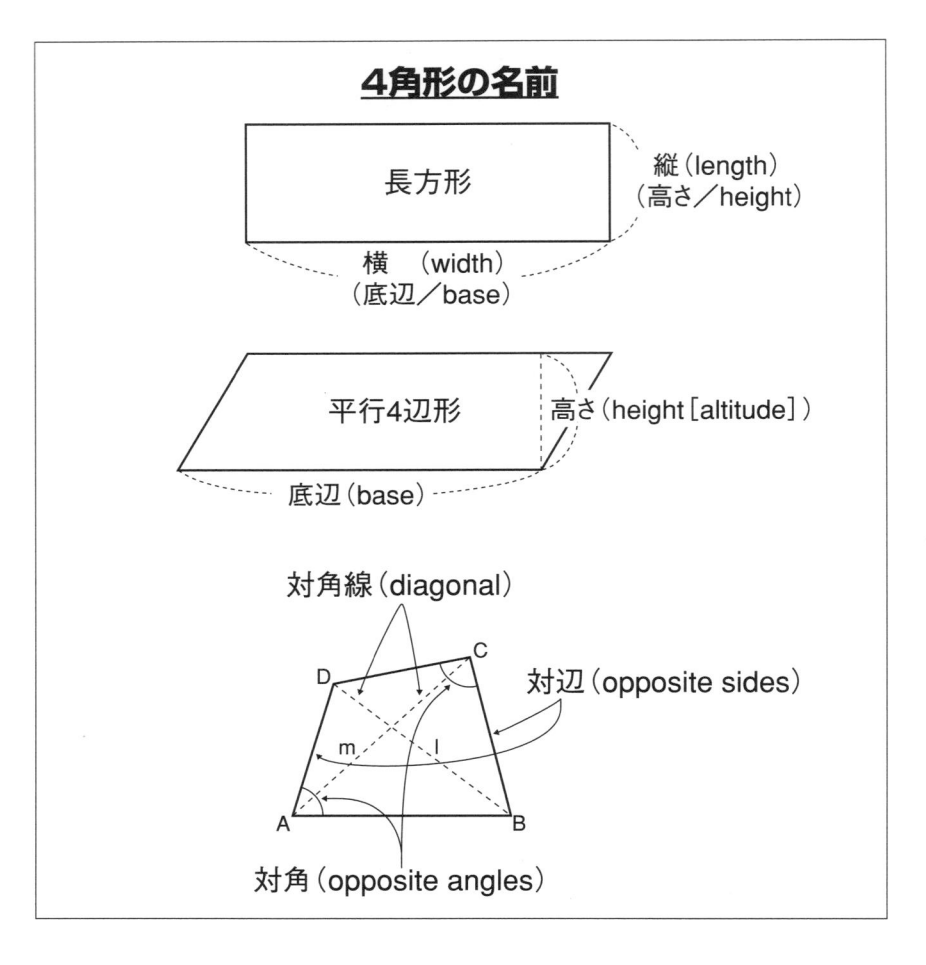

長方形
縦 (length) (高さ／height)
横 (width) (底辺／base)

平行4辺形
高さ (height [altitude])
底辺 (base)

対角線 (diagonal)
対辺 (opposite sides)
対角 (opposite angles)

●4角形の内角の和は360°です。

The sum of the interior angles of a quadrilateral is 360 (three hundred sixty) degrees.

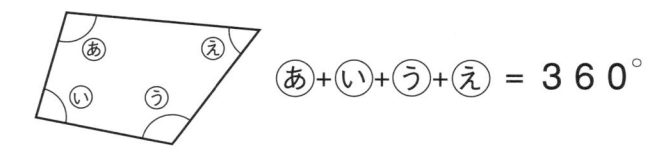

あ＋い＋う＋え ＝ ３６０°

❺ 多角形
Polygons

いろいろな多角形

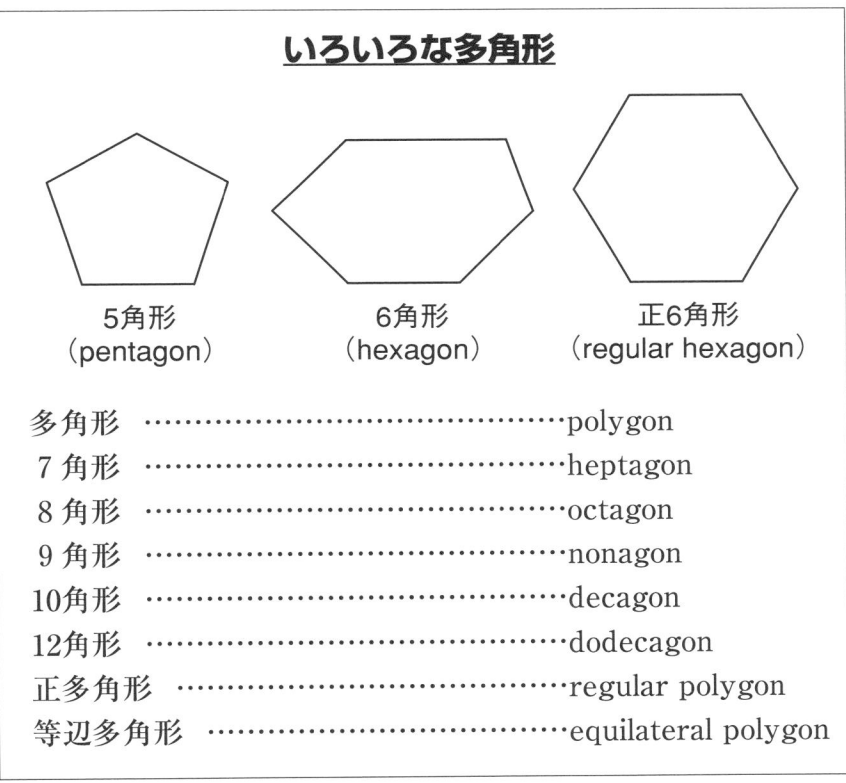

5角形（pentagon）　6角形（hexagon）　正6角形（regular hexagon）

多角形 …………………………………… polygon
7角形 …………………………………… heptagon
8角形 …………………………………… octagon
9角形 …………………………………… nonagon
10角形 ………………………………… decagon
12角形 ………………………………… dodecagon
正多角形 ……………………………… regular polygon
等辺多角形 …………………………… equilateral polygon

多角形と角

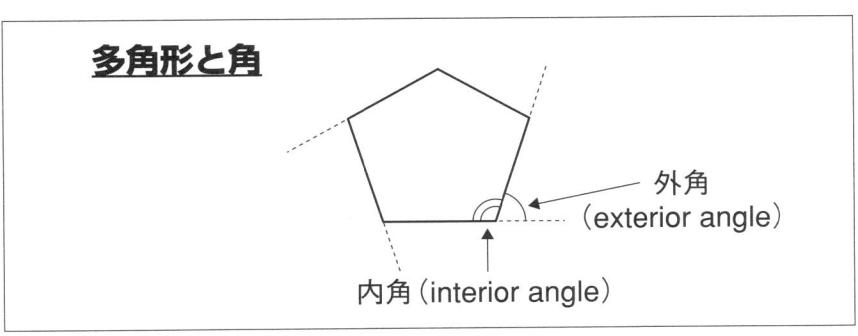

外角（exterior angle）
内角（interior angle）

多角形の名前

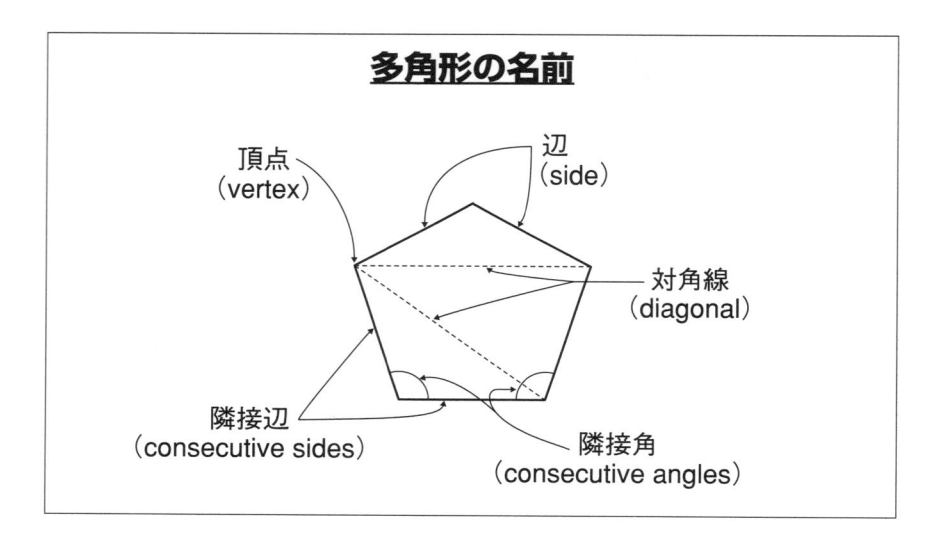

頂点 (vertex)
辺 (side)
対角線 (diagonal)
隣接辺 (consecutive sides)
隣接角 (consecutive angles)

● 5本の直線で囲まれた形を5角形といい，角が5つあります。

A figure bounded by five straight lines is called a pentagon and has five angles.

● 辺の長さがみな等しく，角の大きさもみな等しい形を正多角形といいます。

A figure in which all sides are the same length and all angles also are the same size is called a regular polygon.

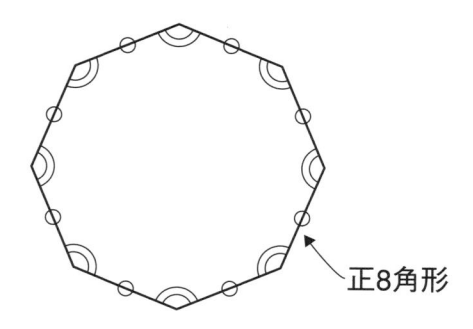

正8角形

● 正6角形を対角線で区切ると，正3角形が6個になります。

Dividing a regular hexagon by its diagonals, we get six equilateral triangles.

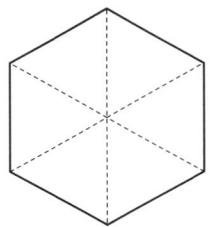

● 多角形の1つの角の内角と外角の和は180°です。

The sum of the interior and exterior angles of one angle in a polygon is 180 (one hundred eighty) degrees.

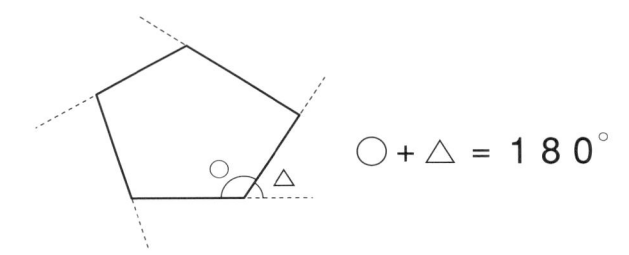

$$\bigcirc + \triangle = 180°$$

● 3角形の角度の和をつぎつぎにたしていけば，多角形の内角の和が求められます。

Adding the sum of the angles in a triangle one after another gives us the sum of the interior angles in a polygon.

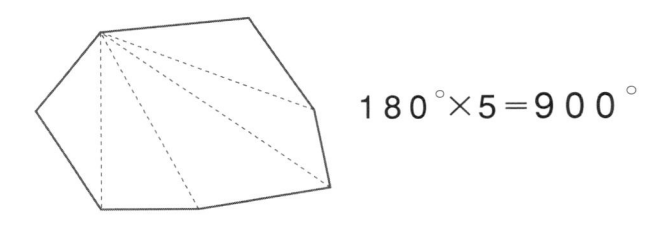

$$180° \times 5 = 900°$$

❻ 合同・相似と角
Congruence, Similarity and Angles

主要なことば

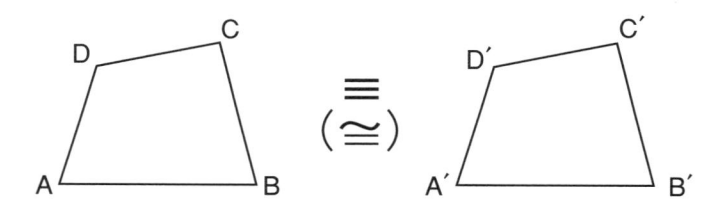

合同	congruence
合同な図形	congruent figures (shapes)
□と○は合同である	□ is congruent to ○.
相似	similarity
相似な図形	similar figures (shapes)
□は○に相似である	□ is similar to ○.
相似比	scale factor
拡大・縮小	enlargement・reduction

対応する角と辺

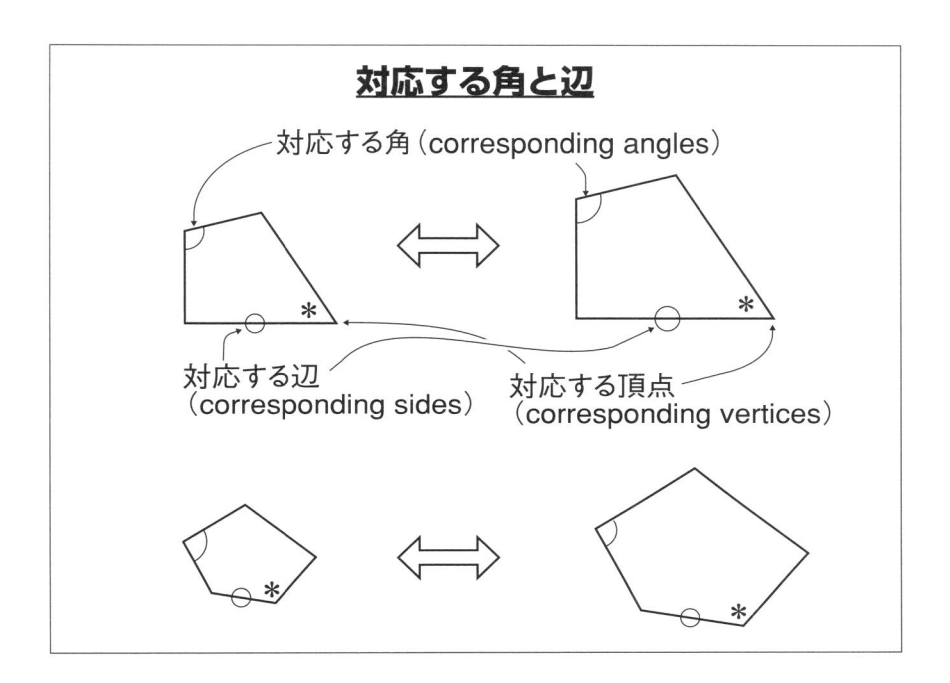

対応する角 (corresponding angles)
対応する辺 (corresponding sides)
対応する頂点 (corresponding vertices)

●2つの図形がまったく同じ形でピッタリ重なりあうとき，2つの図形は合同であるといいます。

When two figures have the same form and overlap each other completely, they are said to be congruent.

●合同な図形において重なる頂点・辺・角をそれぞれ「対応する頂点・辺・角」といいます。

In congruent figures, overlapping vertices, sides and angles are called corresponding vertices, sides and angles, respectively.

●合同な図形は対応する角の大きさと，対応する辺の長さが等しいです。

In congruent figures, the corresponding angles are the same size and the corresponding sides are the same length.

●2つの3角形において，3つの辺がそれぞれ等しいとき，この2つの3角形は合同です（3辺合同）。

Two triangles are congruent when their three sides are equal respectively (S.S.S／side side side).

⇦ 合同 ⇨
（congruent）

●2つの3角形において，2つの辺と，それらに挟まれた1つの角がそれぞれ等しいとき，この2つの3角形は合同です（2辺挟角）。

Two triangles are congruent when two of their sides and the angle between them are equal, respectively (S.A.S／side angle side).

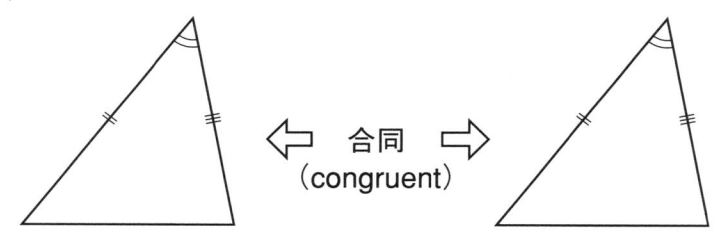

⇦ 合同 ⇨
（congruent）

●2つの3角形において，2つの角と，それらに挟まれた1つの辺がそれぞれ等しいとき，この2つの3角形は合同です（2角挟辺）。

Two triangles are congruent when two of their angles and the side between them are equal, respectively (A.S.A／angle side angle).

⇦ 合同 ⇨
（congruent）

●2つの直角3角形において，斜辺と対応する1辺がそれぞれ等しいとき，2つは合同です。

　Two right triangles are congruent when their hypotenuse and one corresponding side are equal, respectively.

●2つの3角形において，対応する2つの角がそれぞれ等しければ，それらは相似です。

　Two triangles are similar when two corresponding angles are equal, respectively.

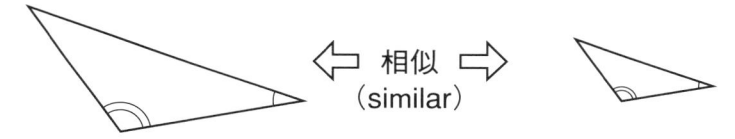

●どんな3角形でも，3つの角の和は180°です。

　In any triangle the sum of three angles is 180 (one hundred eighty) degrees.

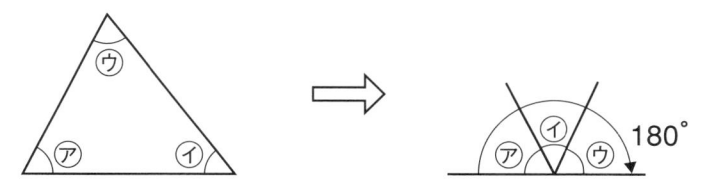

●どんな4角形でも，4つの角の和は360°です。

　In any quadrilateral the sum of four angles is 360 (three hundred sixty) degrees.

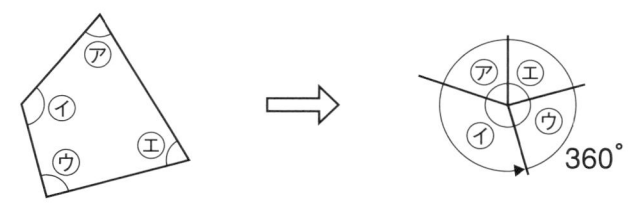

第4章●平面図形と空間図形／ Plane Figures and Solid Figures

⑦ 線対称と点対称
Line Symmetry and Point Symmetry

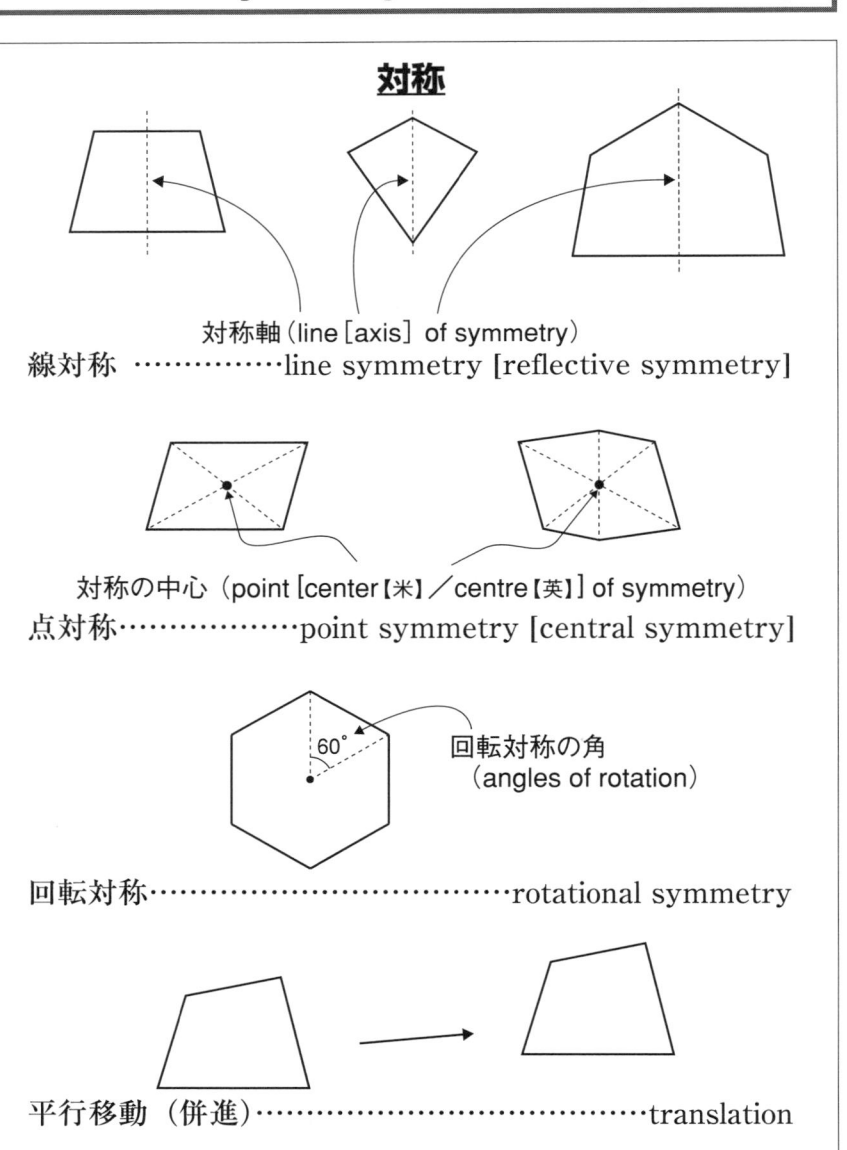

対称

対称軸（line [axis] of symmetry）
線対称 ……………… line symmetry [reflective symmetry]

対称の中心（point [center【米】／centre【英】] of symmetry）
点対称 ……………… point symmetry [central symmetry]

60°
回転対称の角
（angles of rotation）

回転対称 ……………………………… rotational symmetry

平行移動（併進）……………………………… translation

●図形Aは線対称です。図形Bは点対称です。

The figure A has line symmetry. The figure B has point symmetry.

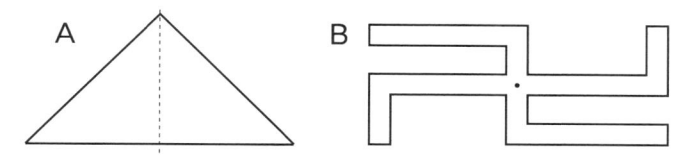

●線対称な図形では，対応する点を結ぶ直線は対称軸と垂直に交わります。また，これらの点から対応する軸までの長さは等しいです。

In a figure having line symmetry, straight lines which link corresponding points are perpendicular to the axis of symmetry. Also, the distance between these points and the corresponding axis of symmetry are equal.

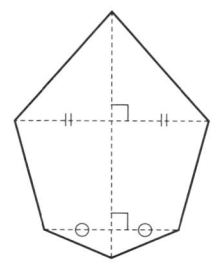

●点対称な図形では，対応する点どうしを結ぶと，かならず対称の中心を通ります。また，点対称な図形では，対称の中心から対応する点までの長さは等しくなっています。

In a figure having point symmetry, lines linking the corresponding points always pass through the center of symmetry. Also, the distance between the point of symmetry and corresponding points are equal.

第4章●平面図形と空間図形／Plane Figures and Solid Figures

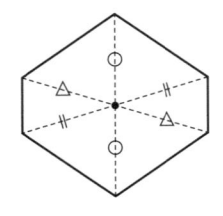

●つぎの図形は線対称か点対称かをそれぞれ答えてください。

Answer whether the following figures have line symmetry or point symmetry, respectively.

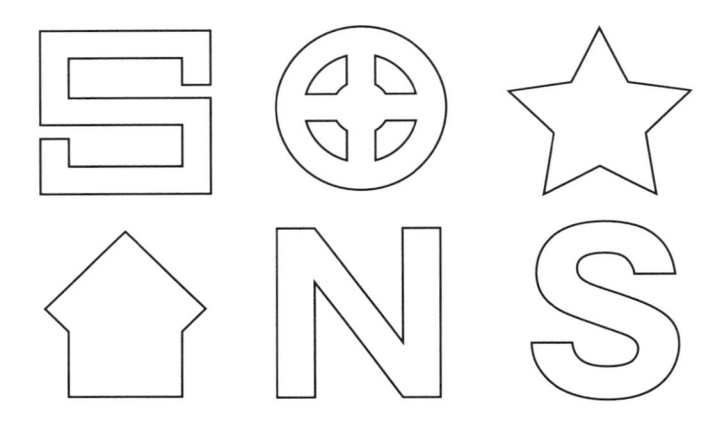

●対称な図形は，対称軸や対称の中心がもつ特徴を利用すると，容易に描けます。

A symmetrical figure can be drawn easily if we use the characteristics of line or point symmetry.

❽ 円と円周

Circles and Circumference

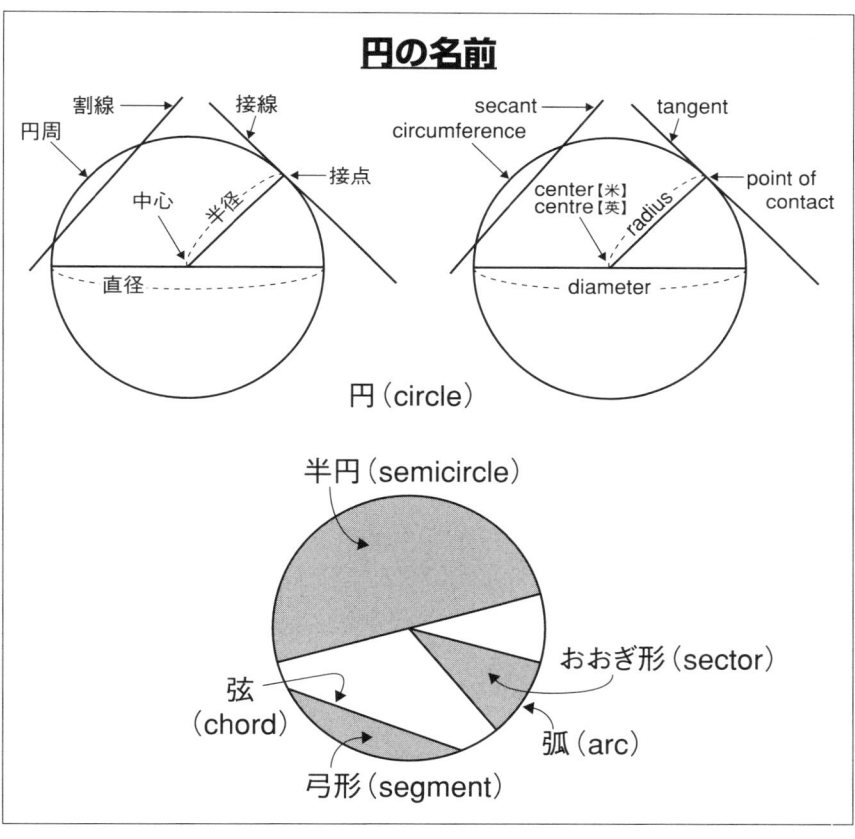

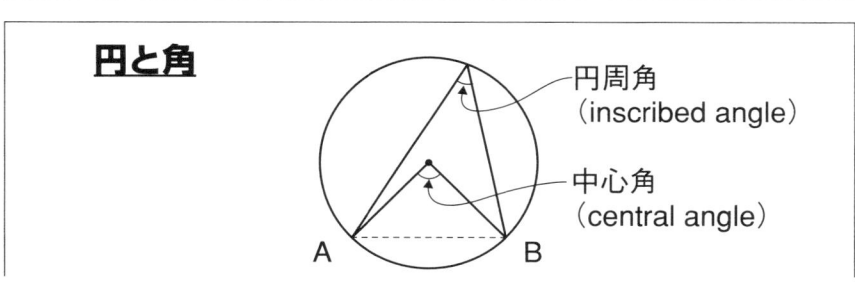

円の名前

割線 / 接線 / 円周 / 中心 / 半径 / 接点 / 直径

secant / tangent / circumference / center【米】centre【英】/ radius / point of contact / diameter

円 (circle)

半円 (semicircle)
おおぎ形 (sector)
弦 (chord)
弧 (arc)
弓形 (segment)

円と角

円周角 (inscribed angle)
中心角 (central angle)

第4章●平面図形と空間図形／ Plane Figures and Solid Figures

内接円（inscribed circle）　　外接円（circumscribed circle）

●直径の長さは半径の2倍です。

The length of the diameter is twice the radius.

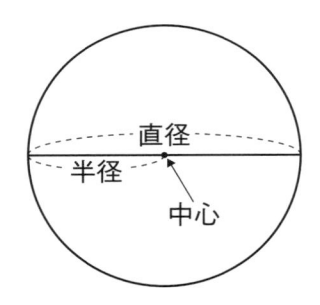

直径
半径
中心

●円周率は3.14として計算しましょう。

Let's calculate using 3.14 (three point one four) for π (pi).

●半径5cmの円を描いてください。

Draw a circle with a radius of 5 cm (five centimeters).

●円周は直径×円周率で求められます。

We can find the circumference by the diameter times π (pi).

●円周は2πrです。

The circumference is equal to $2\pi r$ (two pi r).

⑧―円と円周／Circles and Circumference

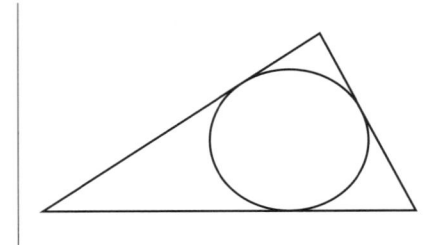

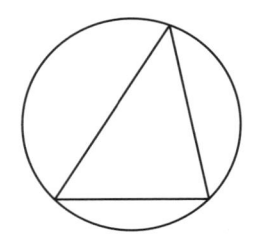

❾ 立方体・直方体と展開図
Cubes, Cuboids and Nets

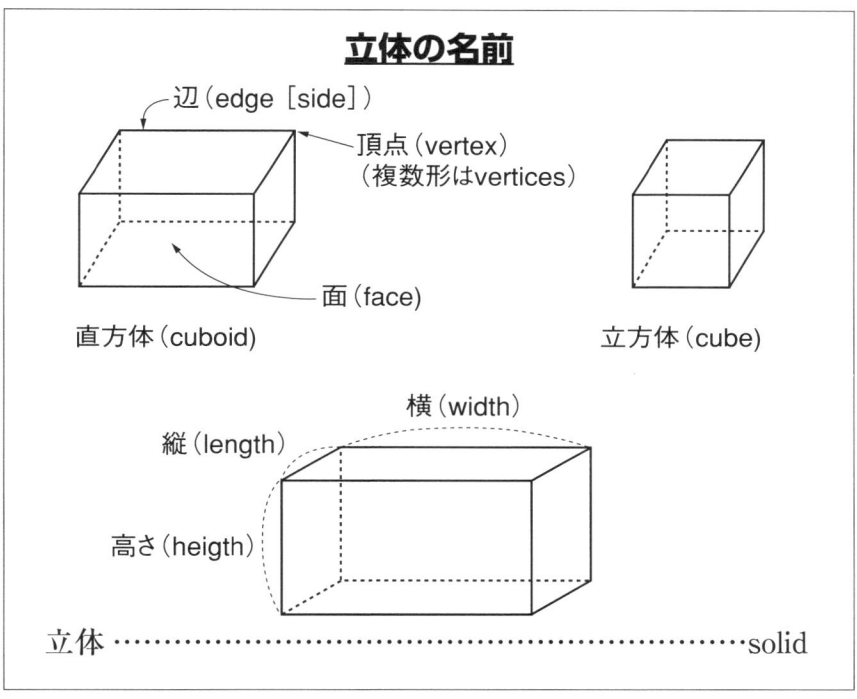

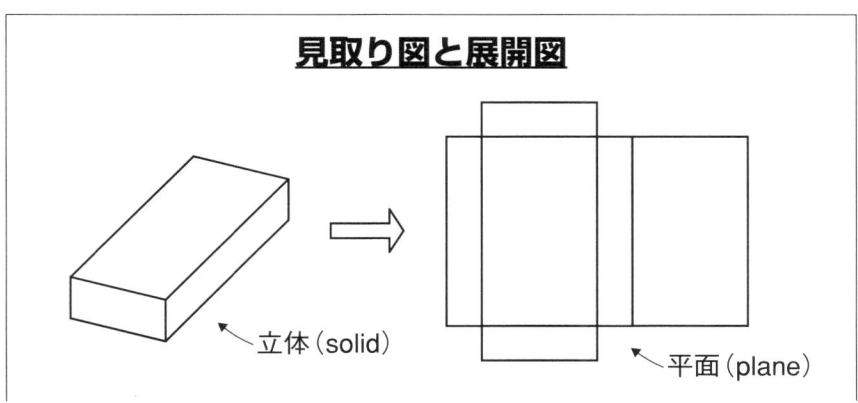

立体の名前

- 辺（edge [side]）
- 頂点（vertex）（複数形はvertices）
- 面（face）

直方体（cuboid）　　　立方体（cube）

- 横（width）
- 縦（length）
- 高さ（heigth）

立体 ……………………………………………………… solid

見取り図と展開図

立体（solid） ⇒ 平面（plane）

第4章●平面図形と空間図形／Plane Figures and Solid Figures

見取り図(sketch)　　　　　　　展開図(net)

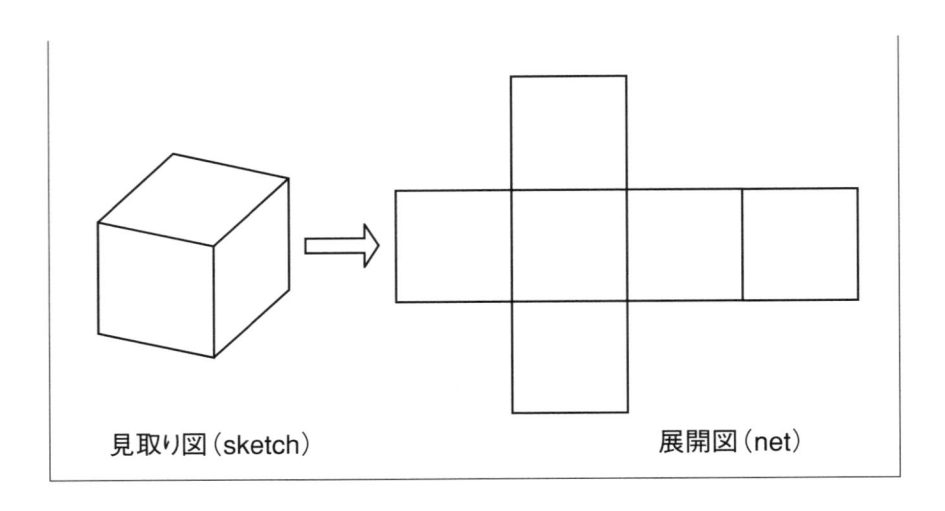

●直方体や立方体では，向かいあっている面は平行，隣りあっている面は垂直です。

In a cuboid or cube, faces facing each other are parallel and faces adjacent to each other are perpendicular.

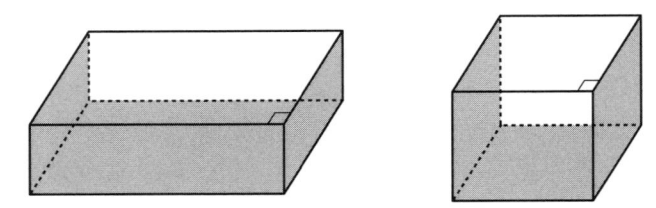

●見ただけで全体のおおよその形がわかる図を見取り図といいます。

A figure through which we can understand an overall rough form just by looking at it is called a sketch.

●立体を辺にそって切り開き，平面上に書いた図を展開図といいます。

A figure which we draw on a plane by opening a solid along its sides is called a net.

● 1辺が10cmの正方形の紙を6枚あわせると，1辺が10cmの立方体を作ることができます。

Combining six pieces of square paper with a side length of 10 cm (ten centimeters), we get a cube whose side is 10 cm (ten centimeters).

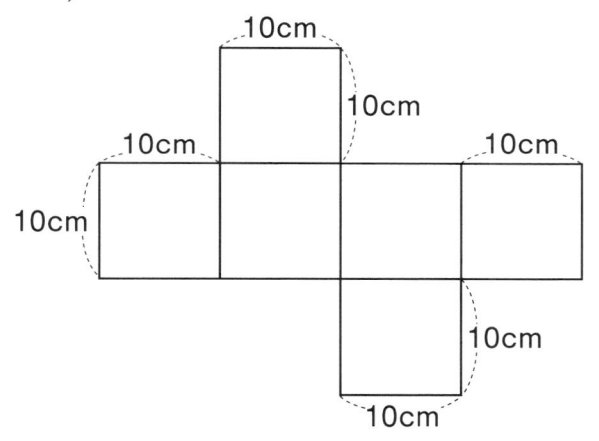

● 下図において辺ABと辺CDと辺EFと辺GHは平行，辺ABと辺BCと辺BFは垂直になっています。

In the following figure, side AB, CD, EF, and GH are parallel, and side AB, BC, and BF are perpendicular.

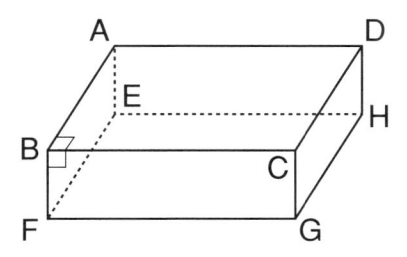

● 展開図を折っていくと，立体ができあがります。

Folding a net, we get a complete solid.

第4章● 平面図形と空間図形／ Plane Figures and Solid Figures   119

●1辺が15cmの立方体

a cube which has side of 15 cm (fifteen centimeters)

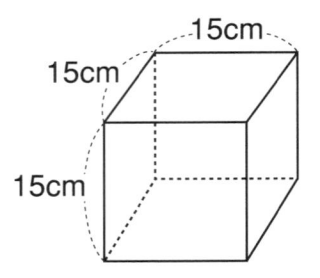

●直方体の大きさは縦の長さ，横の長さ，高さで表わします。

The size of a cuboid is expressed by its length, width and height.

●立方体はどの辺も長さが同じなので，大きさは1辺の長さで表わします。

The size of a cube is expressed by its side, because all sides have the same length.

⑨―立方体・直方体と展開図／Cubes, Cuboids and Nets

⑩ 柱と錐と球
Prisms, Pyramids and Spheres

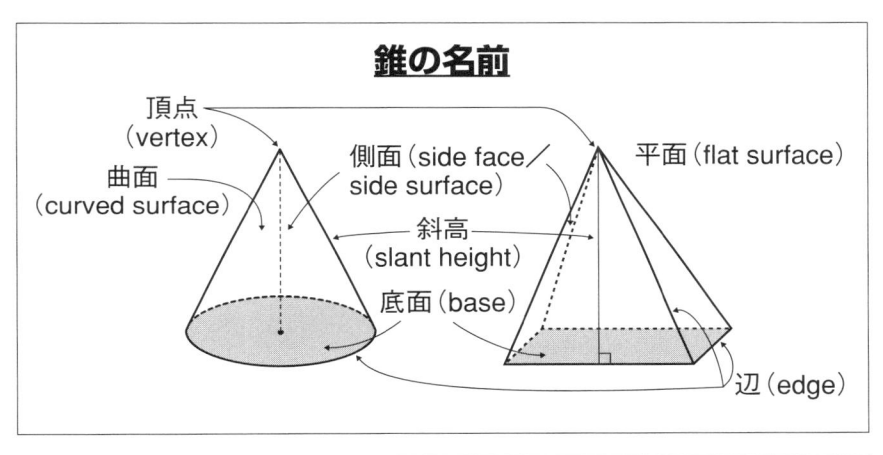

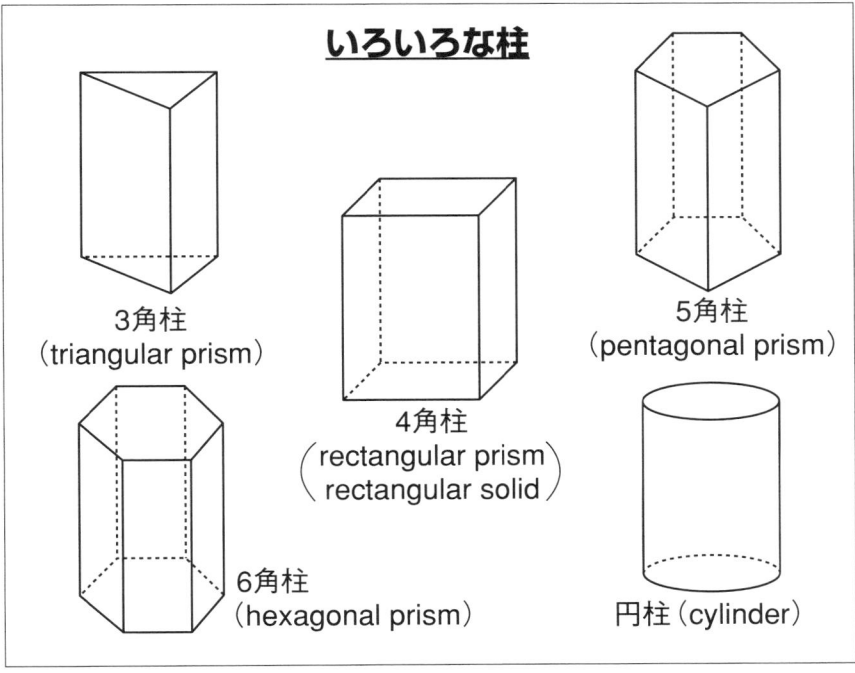

錐の名前

- 頂点 (vertex)
- 曲面 (curved surface)
- 側面 (side face / side surface)
- 斜高 (slant height)
- 底面 (base)
- 平面 (flat surface)
- 辺 (edge)

いろいろな柱

- 3角柱 (triangular prism)
- 4角柱 (rectangular prism / rectangular solid)
- 5角柱 (pentagonal prism)
- 6角柱 (hexagonal prism)
- 円柱 (cylinder)

第4章●平面図形と空間図形／Plane Figures and Solid Figures

いろいろな錐

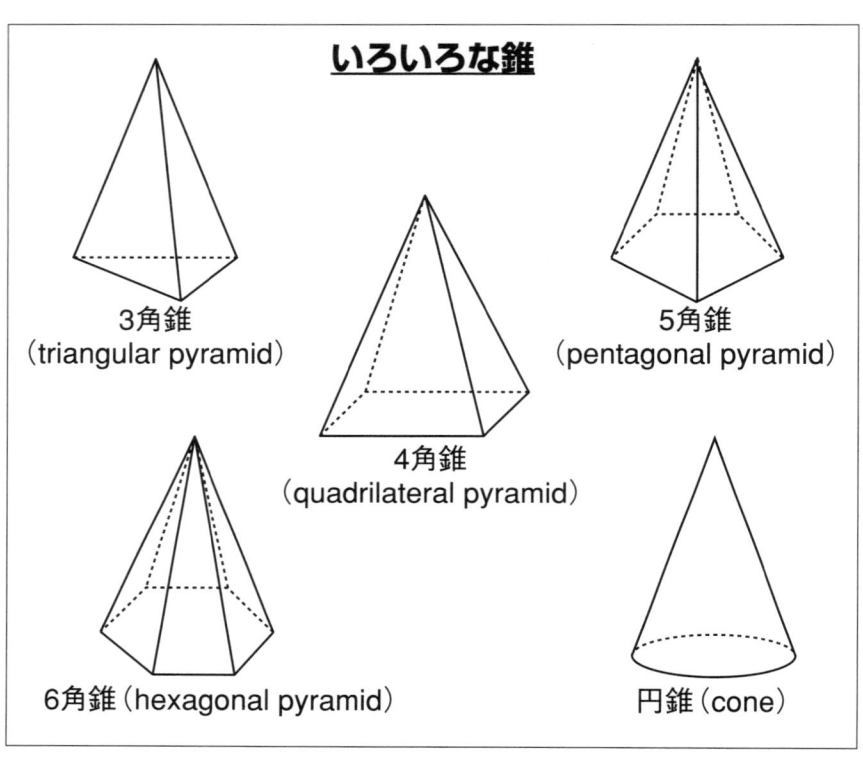

3角錐（triangular pyramid）

4角錐（quadrilateral pyramid）

5角錐（pentagonal pyramid）

6角錐（hexagonal pyramid）

円錐（cone）

球の名前

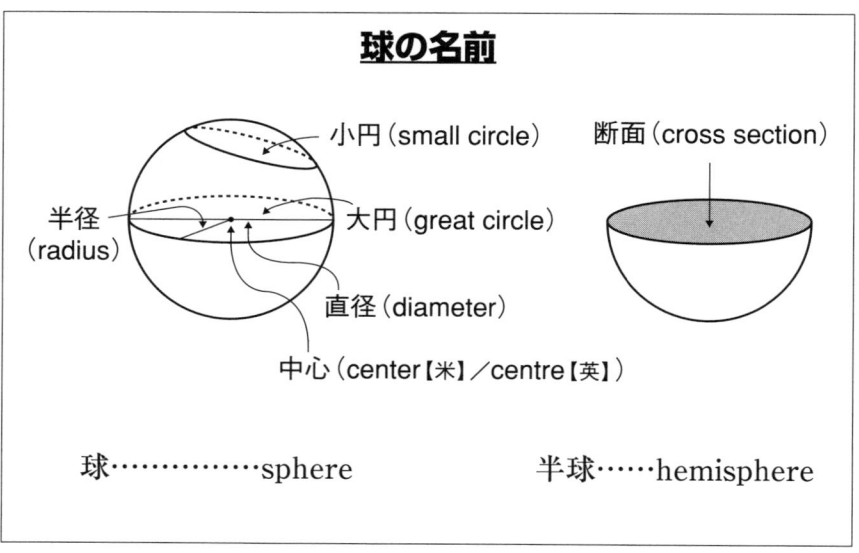

小円（small circle）
半径（radius）
大円（great circle）
直径（diameter）
中心（center【米】／centre【英】）
断面（cross section）

球……sphere　　　　半球……hemisphere

展開図

4角柱（rectangular prism）　　円柱（cylinder）

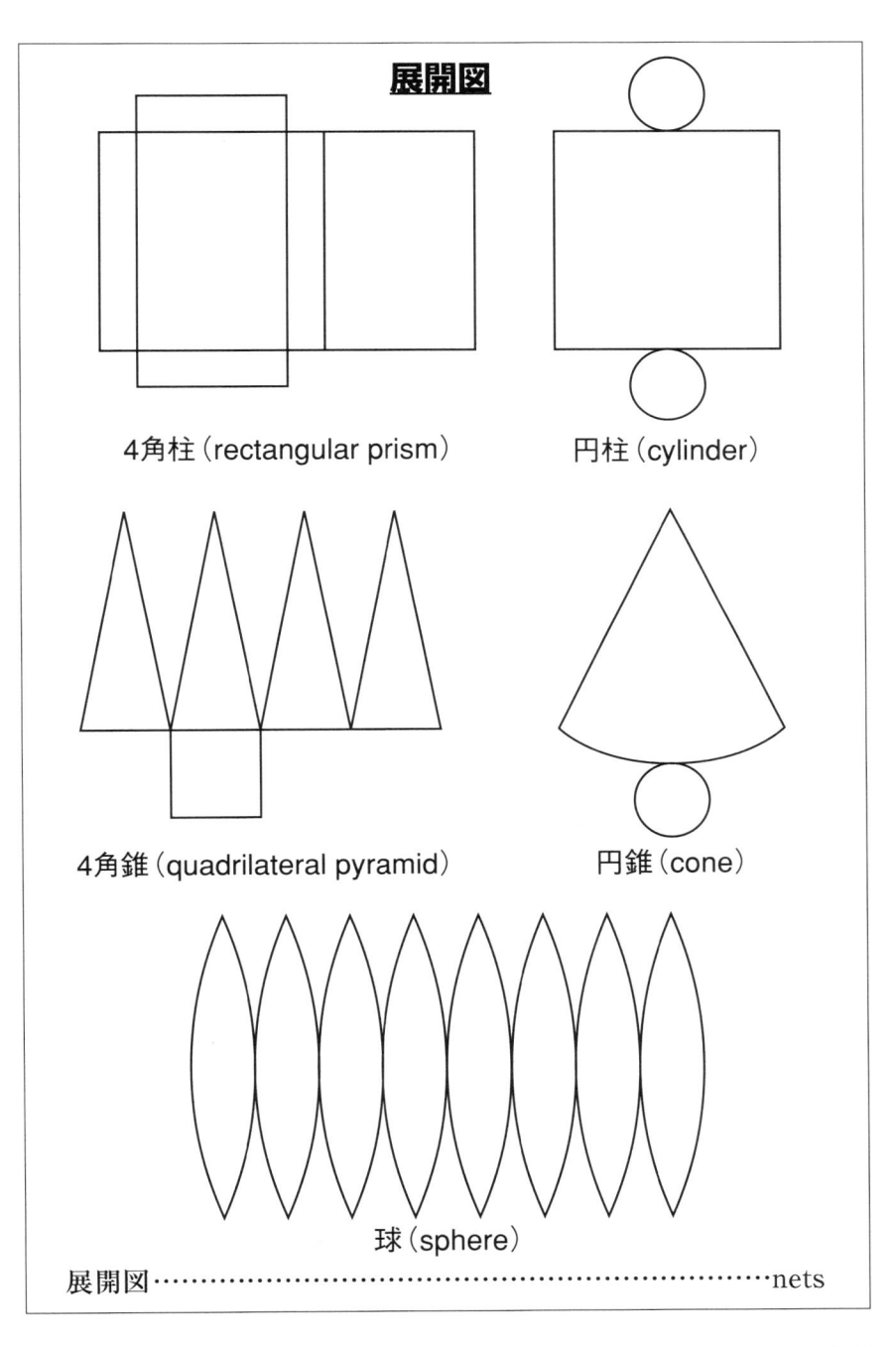

4角錐（quadrilateral pyramid）　　円錐（cone）

球（sphere）

展開図 …………………………………………………… nets

第4章●平面図形と空間図形／Plane Figures and Solid Figures

●立方体や直方体は4角柱といえます。

A cube or cuboid can be called a rectangular prism.

●正角錐では底面が正多角形，側面が2等辺3角形になっています。

In a right pyramid the base is a regular polygon and the side faces are isosceles triangles.

●円錐の底面は円，側面は曲面になっています。

In a cone the base is a circle and the side face is a curved surface.

●角錐や円錐では頂点から底面に垂直にひいた直線の長さを高さといいます。

In a pyramid or cone, the length of a straight line drawn from the top vertex and perpendicular to the base is called the height.

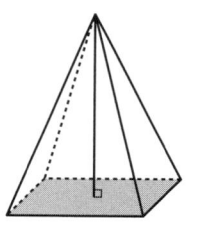

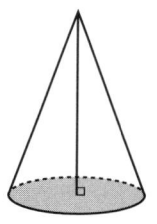

●正5角柱には5角形の面が2つ，長方形の面が5つ，頂点が10個，辺が15個あります。

A right pentagonal prism has 2 (two) pentagonal faces, 5 (five) rectangular faces, 10 (ten) vertices, and 15 (fifteen) sides.

●頂点の数は3角柱なら3の2倍，4角柱なら4の2倍です。

The number of the vertices of a triangular prism is twice the number 3 (three), and that of a rectangular prism is twice the number 4 (four).

●つぎのような円錐があります。展開図を書きましょう。
Draw the net for a cone like the following.

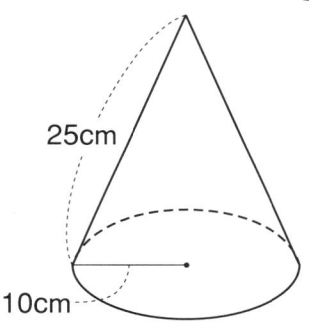

25cm
10cm

●立体では真正面と真上から見た形を組合せて表わすことがよくあります。
A solid is often expressed by the combination of a figure showing the view from front and a figure showing the view from above.

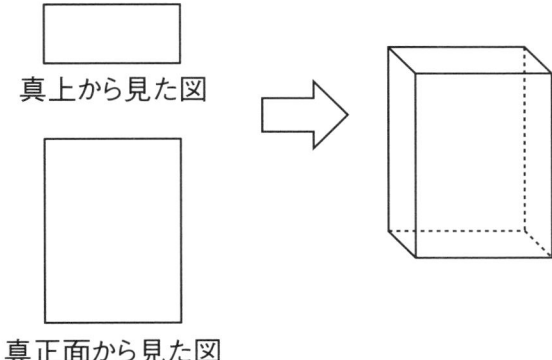

真上から見た図

真正面から見た図

●球はどこから見ても，かならず円に見えます。
A sphere is always seen as a circle from wherever we look at it.

チャレンジ・テスト④ CHALLENGE 練習問題

① Which straight lines are parallel to each other, and which lines are perpendicular to each other?
［平行な直線，垂直な直線はそれぞれどれとどれですか。］

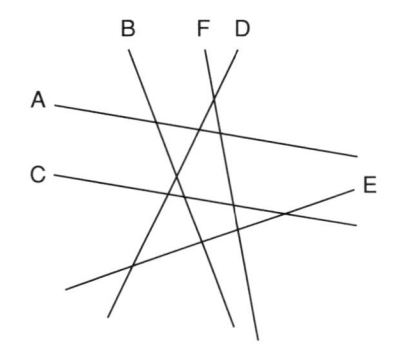

② Line A is parallel to line B, and line C is parallel to line D. Find the angle measures of a, b, c, and d.
［直線AとB，CとDは平行です。a，b，c，dの角度を求めなさい。］

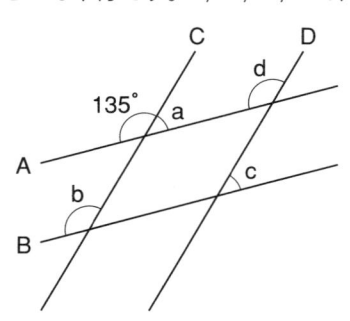

③ How many degrees do angles a, b, c, d, and e have respectively, in the following figure?
［つぎの図を見て，∠a，∠b，∠c，∠d，∠eはそれぞれ何度ですか。］

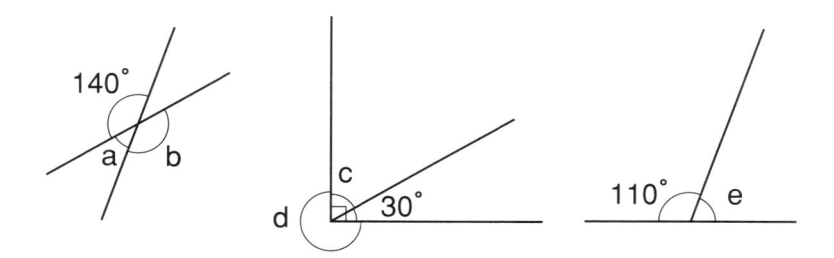

④ Two triangle rulers are arranged together as shown below. What are the sizes of the angles a, b and c respectively?

［三角定規がつぎのように組合さっています。a，b，cの角度はそれぞれいくらですか。］

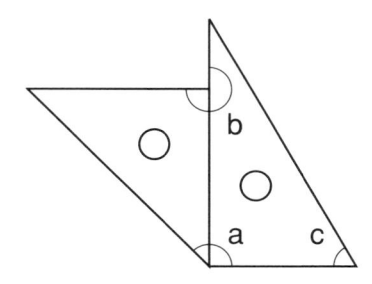

⑤ In the triangle shown below, AD is the bisector of angle A. How many degrees is angle a?

［つぎの３角形ABCで，ADは∠Aの２等分線です。∠aは何度ですか。］

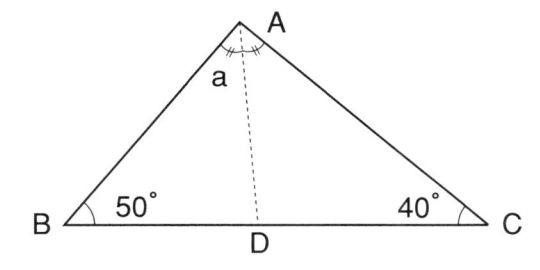

⑥ Triangle ABD is an equilateral triangle, and triangle BCD is an isosceles triangle. How long is side AC?

[3角形ABDは正3角形，3角形BCDは2等辺3角形です。辺ACの長さはいくらですか。]

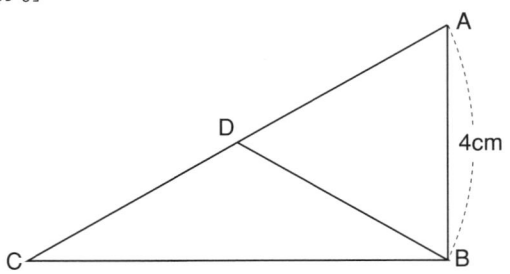

⑦ Name the kinds of quadrilateral of a pair of parallel opposite sides.

[向かい合う1組の辺が平行な4角形にはどんな図形がありますか。]

⑧ A shape is formed by combining a parallelogram, a square, an equilateral triangle, and a trapezoid. How many degrees are angles a, b, c, and d, respectively?

[平行4辺形，正方形，正3角形，台形を組合せて形を作りました。∠a，∠b，∠c，∠dはそれぞれ何度ですか。]

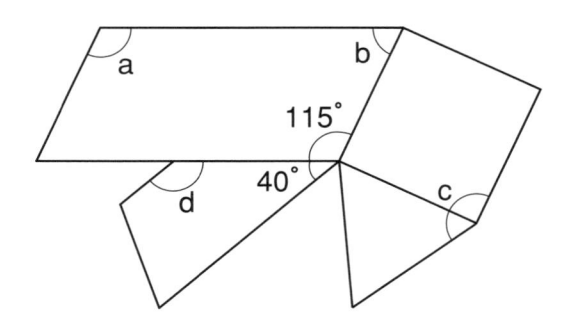

⑨ How many degrees does the central angle of a regular pentagon have? Also, how many diagonals does a hexagon have?

［正5角形の中心角は何度ですか。また，6角形には対角線が何本ありますか。］

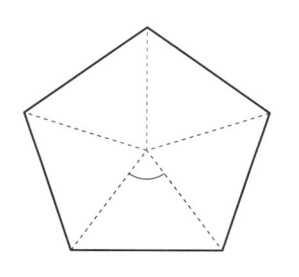

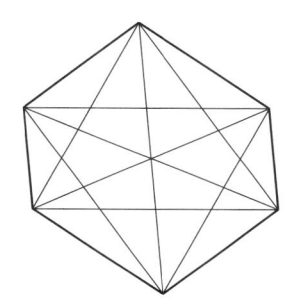

⑩ Among the quadrilaterals shown below, which is the one where two congruent triangles can be made when divided by a diagonal?

［つぎの4角形の中で，1本の対角線でわけると，合同な3角形ができるのはどれですか。］

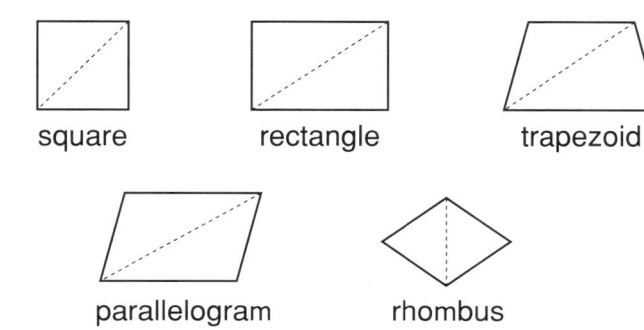

square   rectangle   trapezoid

parallelogram   rhombus

⑪ Figure A is similar to figure B as shown below. How many degrees do angles e, f, g, and h have respectively?

[つぎのAとBは相似な図形です。∠e, ∠f, ∠g, ∠h, はそれぞれ何度ですか。]

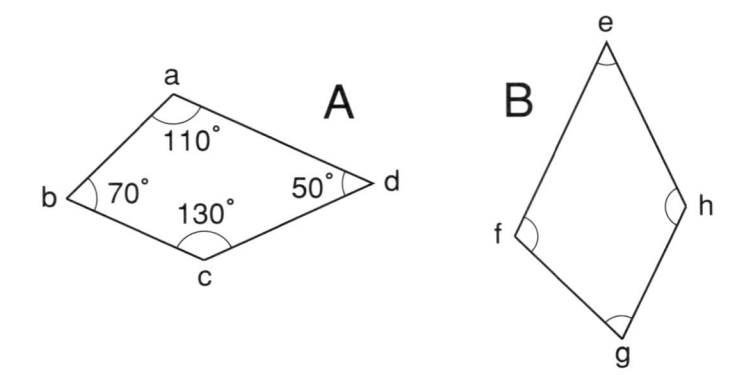

⑫ Does each of the following figures have line symmetry or point symmetry?

[つぎの図は線対称ですか。それとも点対称ですか。]

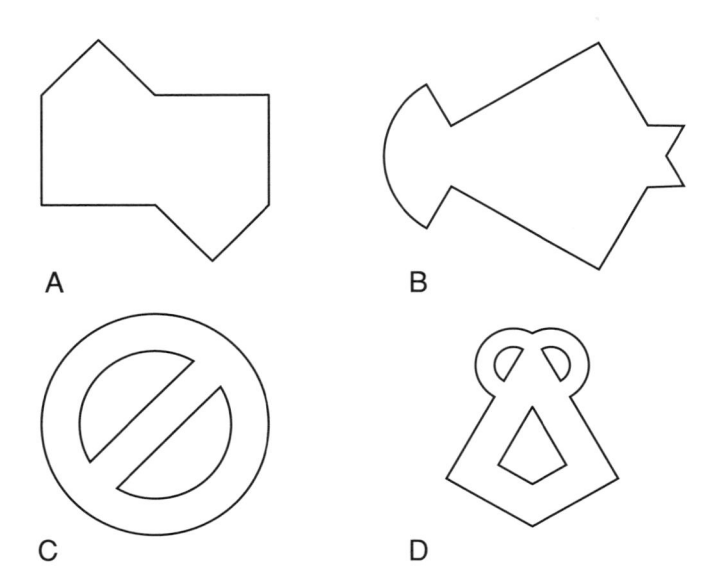

⑬ The following figure has point symmetry. Where is the point (center) of symmetry?

［つぎの図は点対称な図形です。対称の中心はどこですか。］

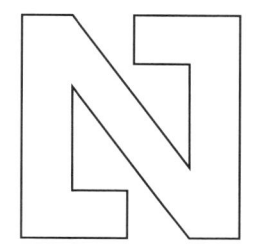

⑭ What is the diameter of circle C shown in the following figure?

［つぎの図において円Cの直径は何cmですか。］

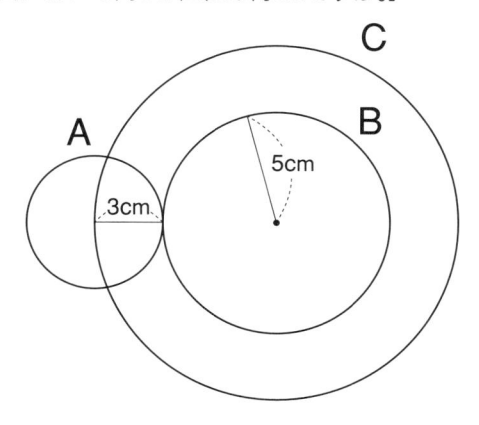

⑮ What is the circumference of a circular pond which has a radius of 7 m (seven meters)? Calculate using 3.14 (three point one four) for π (pi).

［半径7mの池の周囲は何mになりますか。円周率は3.14として計算しましょう。］

⑯ In order to make a cuboid with a length of 5 cm (five centimeters), a width of 9 cm (nine centimeters), and a height of 3 cm (three centimeters), what kind of rectangles, and how many each, are necessary?

［縦5cm，横9cm，高さ3cmの直方体を作るには，どんな長方形の紙がそれぞれ何枚ずつ必要ですか。］

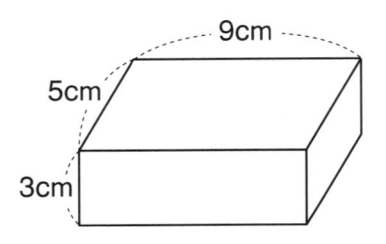

⑰ A net of a cuboid is shown below. What are (1) the point(s) which overlap point b, (2) the edge(s) which overlap edge gh, and (3) the face(s) which are parallel to face B.

［つぎは直方体の展開図です。(1)点 b と重なる点，(2)辺 gh と重なる辺，(3)面 B と平行になる面をそれぞれ答えなさい。］

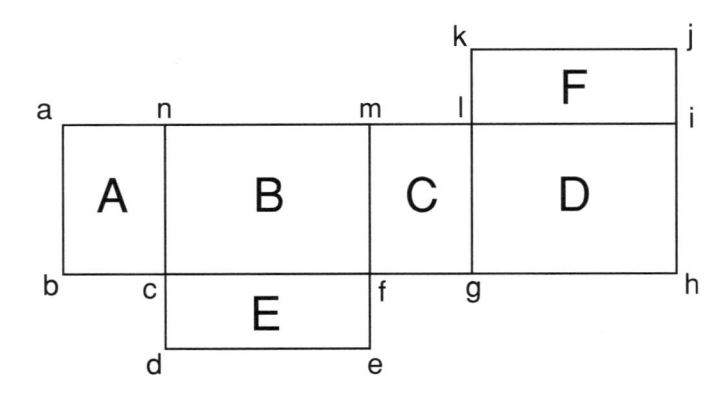

⑱ What kind of figures are the following nets, respectively?

［つぎはどんな図形の展開図でしょうか。］

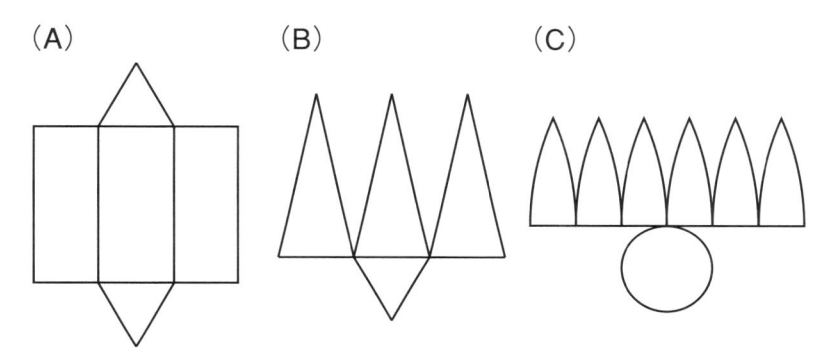

(A)　　　　　(B)　　　　　(C)

⑲ Draw the shapes viewing the following figures from the front and from directly overhead respectively.

［つぎの図の正面と真上から見た形を書きましょう。］

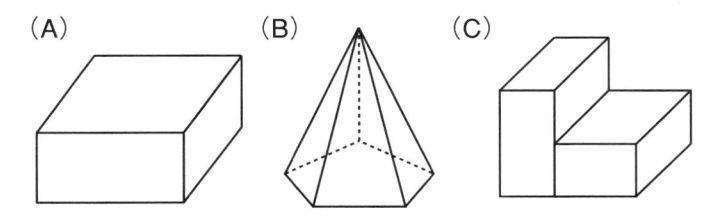

(A)　　　　　(B)　　　　　(C)

年月日・年代・世紀

Tea Room ❷

① − 順は「日→月→年」（イギリス式）もしくは「月→日→年」（アメリカ式）のどちらかで，文中では統一します。
　　24 May 2005／May 24, 2005（2005年5月24日）

② − 7/8/00のように，日づけを数字とスラッシュで表わすことはできるだけ避けます。イギリス人は2000年8月7日，アメリカ人は2000年7月8日と理解してしまいます。ただし，入国カードなどでは，アメリカを含む多くの国で「日→月→年」（イギリス式）の順序で数字で記入します。

③ − 文末の場合を除いて，年の後にはカンマを打ちます。
　　On October 23, 1945, he was born in Kanagawa-ken.
　　　——1945年の10月23日に，彼は神奈川県で生まれました。

④ − 西暦の数字には3桁区切りのカンマはつけません。
　　○　2005　（2005年）
　　×　2,005　（2005年）

⑤ − 「○○年代」は年の後に-sをつけます。アルファベットでも数字でもかまいません。
　　the eighties／the 1980 s（1980年代），the nineties／the 1990s（1990年代）

⑥ − 世紀は数字でもアルファベットでも表わせます。形容詞的用法の場合はcenturyの前にもハイフンを入れます。
　　the twenty-first century（21世紀），a 19th-century building（19世紀の建物）

第5章 面積と体積

Areas and Volumes

① 面積・体積・容積と単位
Areas, Volumes, Capacities and Units

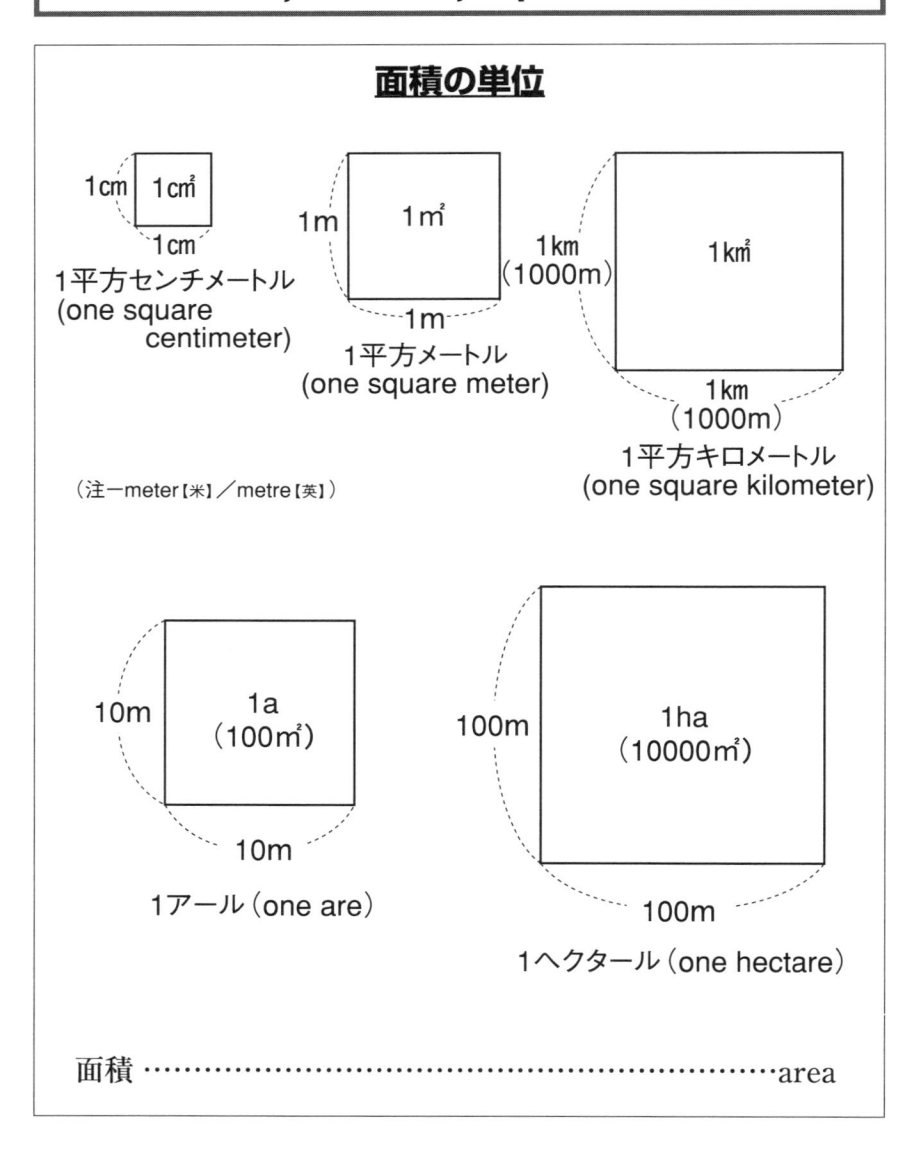

面積の単位

- 1cm × 1cm = 1cm²
  1平方センチメートル (one square centimeter)

- 1m × 1m = 1m²
  1平方メートル (one square meter)

- 1km (1000m) × 1km (1000m) = 1km²
  1平方キロメートル (one square kilometer)

（注―meter【米】／metre【英】）

- 10m × 10m = 1a (100m²)
  1アール (one are)

- 100m × 100m = 1ha (10000m²)
  1ヘクタール (one hectare)

面積 ······ area

体積の単位

1立方センチメートル
(one cubic centimeter)

1立方メートル
(one cubic meter)

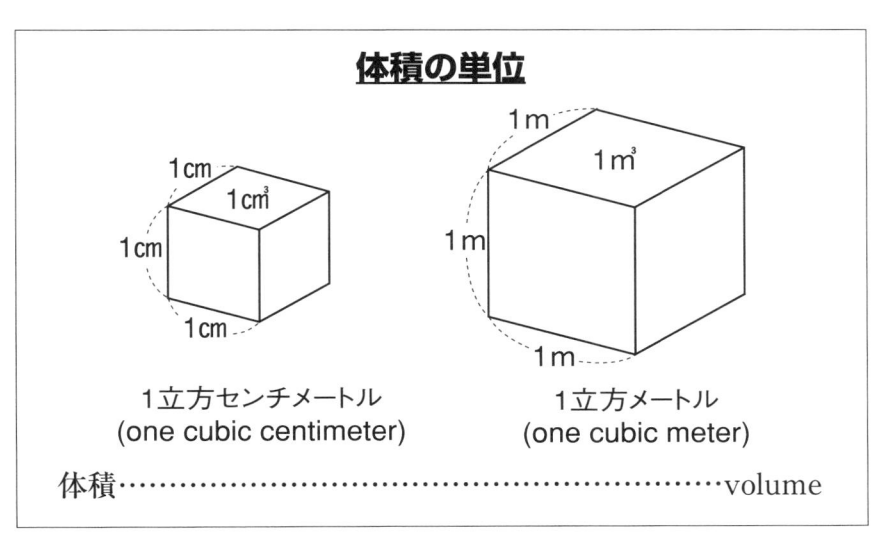

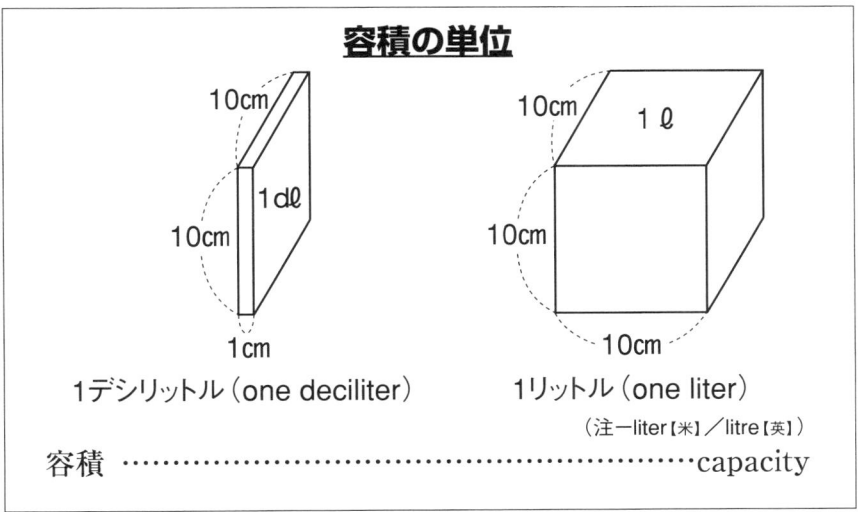

体積 ·· volume

容積の単位

1デシリットル (one deciliter)

1リットル (one liter)
（注―liter【米】／litre【英】）

容積 ·· capacity

●1辺が1cmの正方形の面積を1cm²と書いて1平方センチメートルと読みます。

The area of a square with sides of 1 cm (one centimeter) is written as 1 cm² and read as one square centimeter.

第5章●面積と体積／Areas and Volumes

●1m²は10000 cm²と同じです。

1 m² (One square meter) is the same as 10,000 cm² (ten thousand square centimeters).

●100m²を1ａと書き，1アールといいます。

100 m² (One hundred square meters) is written as 1 a and read as one are.

●10000 m²を1haと書き，1ヘクタールといいます。

10000 m² (Ten thousand square meters) is written as 1 ha and read as one hectare.

●1辺が1kmの正方形の面積は1km²で，1平方キロメートルといいます。

The area of a square where each side is 1 km (one kilometer) is 1 km² and read as one square kilometer.

●1辺が1cmの立方体と同じ体積を1cm³と書き，1立方センチメートルといいます。

The volume of a cube where each side is 1 cm (one centimeter) is written as 1 cm³ and read as one cubic centimeter.

●1辺が1mの立方体と同じ体積を1m³と書き，1立方メートルと読みます。

The volume of a cube where each side is 1 m (one meter) is written as 1 m³ and read as one cubic meter.

●1000 cm³ は1ℓで，1ℓは10dℓ です。

1000 cm³ (One thousand cubic centimeters) is 1 (one) liter, and 1 (one) liter is 10 (ten) deciliters.

● 100 cm³ は1dℓ のことで，水なら100ｇになります。
100 cm³ (One hundred cubic centimeters) is the same as 1 (one) deciliter, and it weighs 100 (one hundred) grams when it is water.

● つぎの平面図形の斜線部の面積を求めましょう。
Find the area of the hatched part of the following plane figure.

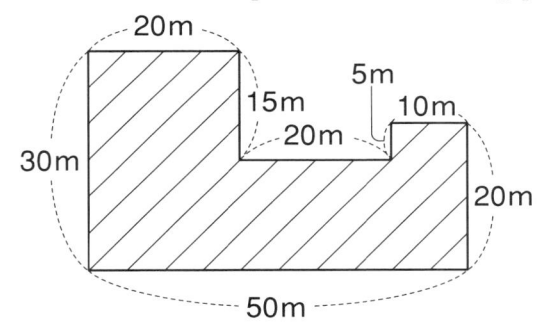

● つぎの立体の体積はいくらでしょうか。
How much is the volume of the following solid figure?

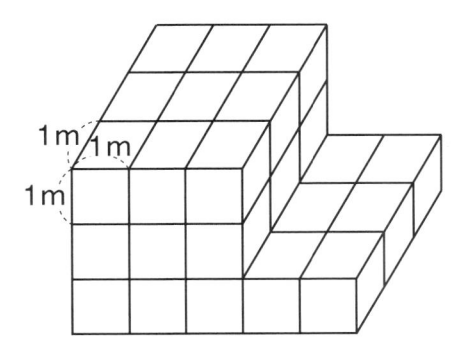

第5章●面積と体積／Areas and Volumes

② 3角形・4角形・多角形と円の面積

Area of Triangles, Quadrilaterals, Polygons and Circles

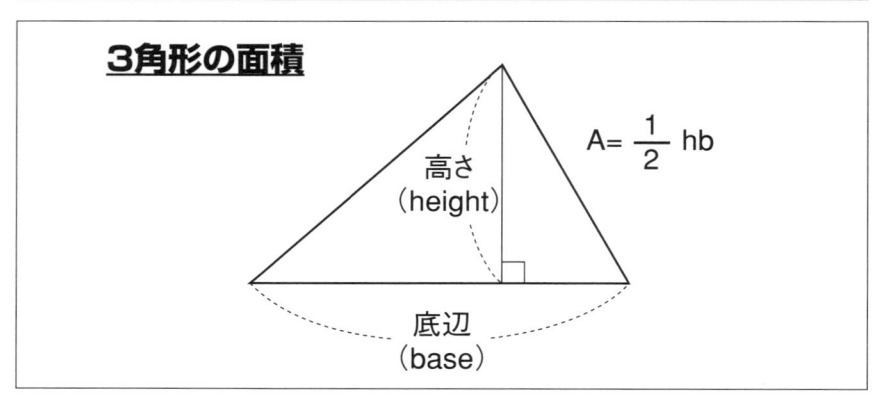

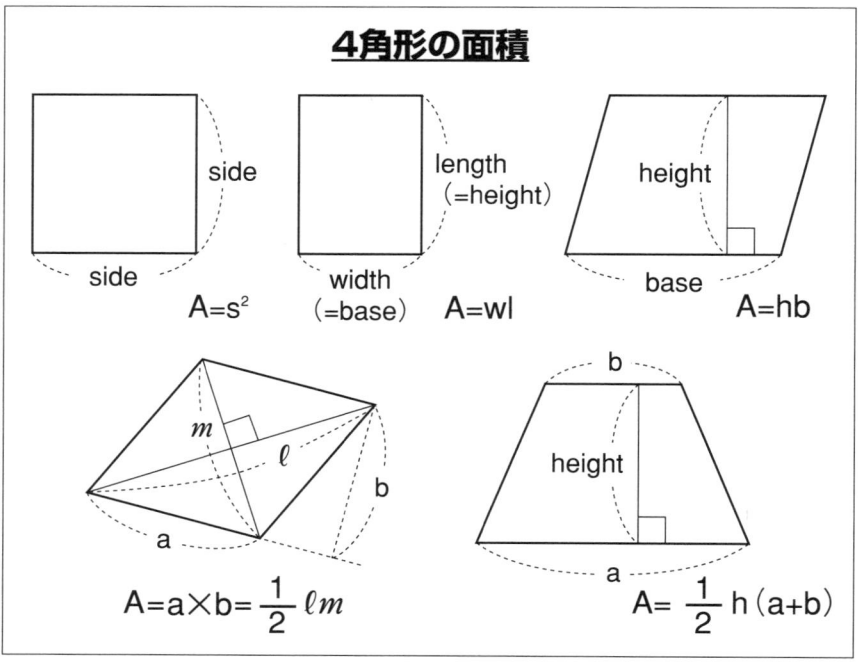

3角形の面積

高さ（height）
底辺（base）
$A = \frac{1}{2} hb$

4角形の面積

side / side
$A = s^2$

length (=height) / width (=base)
$A = wl$

height / base
$A = hb$

m, ℓ, a, b
$A = a \times b = \frac{1}{2} \ell m$

b, height, a
$A = \frac{1}{2} h(a+b)$

円の面積

$A = \pi r^2$

●3角形の面積は，底辺×高さ×$\frac{1}{2}$で求められます。

The area of a triangle can be obtained by multiplying base times height times $\frac{1}{2}$ (one half).

●ひし形の面積を求めるには，底辺×高さで計算します。

To find the area of a rhombus, calculate base times height.

●多角形の面積は，その多角形をいくつかの3角形にわけ，それぞれの3角形の面積をあわせることで求めます。

The area of a polygon is obtained by dividing the polygon into several triangles and summing up the areas of these triangles.

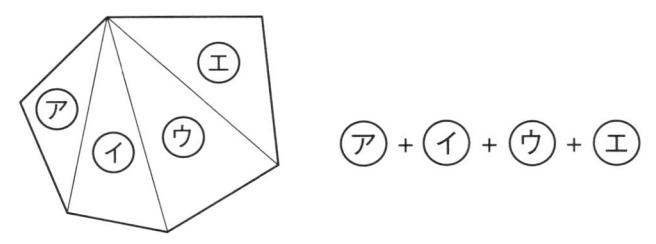

ア + イ + ウ + エ

●円の面積を求める公式は，半径×半径×3.14です。3.14を円周率といいます。

The formula for finding the area of a circle is radius times radius times 3.14 (three point one four). 3.14 (Three point one four) is called pi.

❸ 方体・柱・錐・球の表面積

Surface Areas of Cuboids, Prisms, Pyramids and Spheres

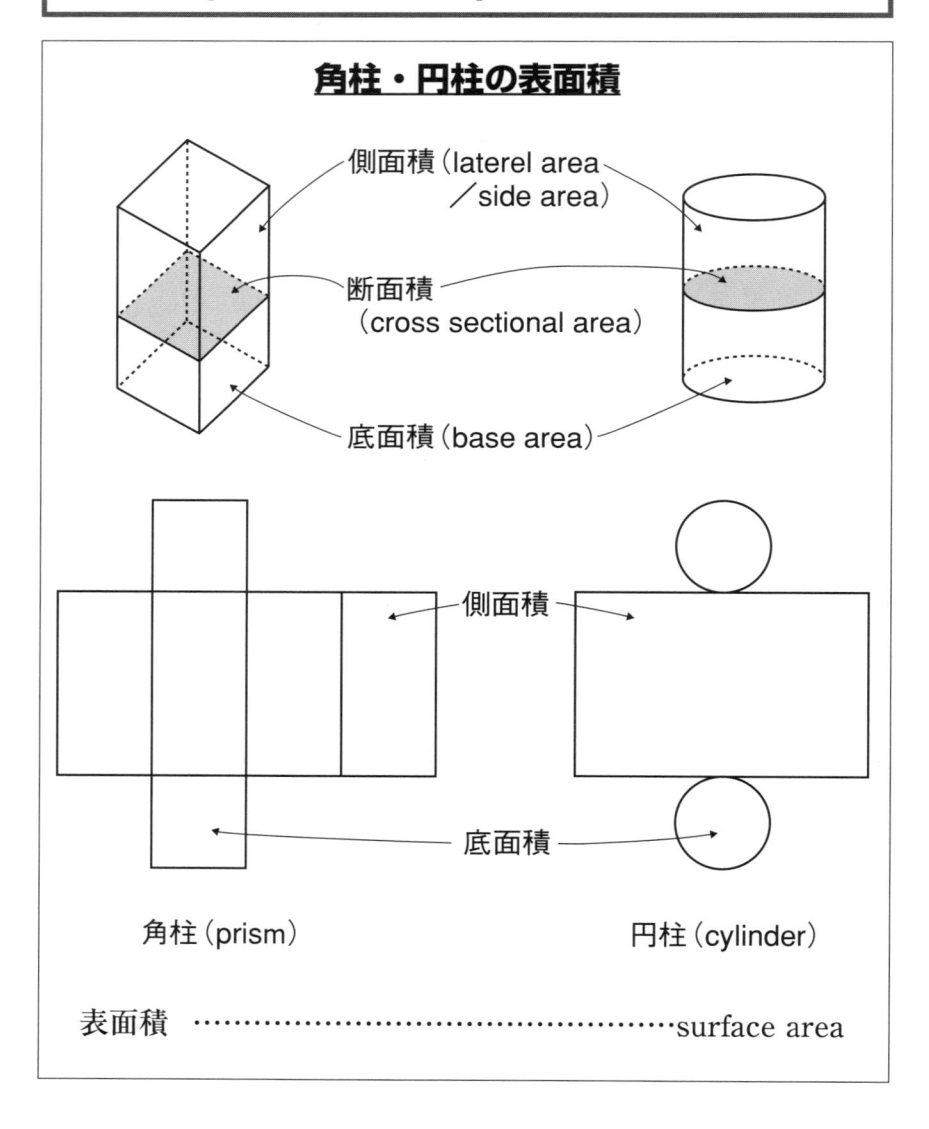

角柱・円柱の表面積

側面積（laterel area／side area）

断面積（cross sectional area）

底面積（base area）

角柱（prism）

円柱（cylinder）

側面積

底面積

表面積 ……………………………… surface area

3角錐・円錐の表面積

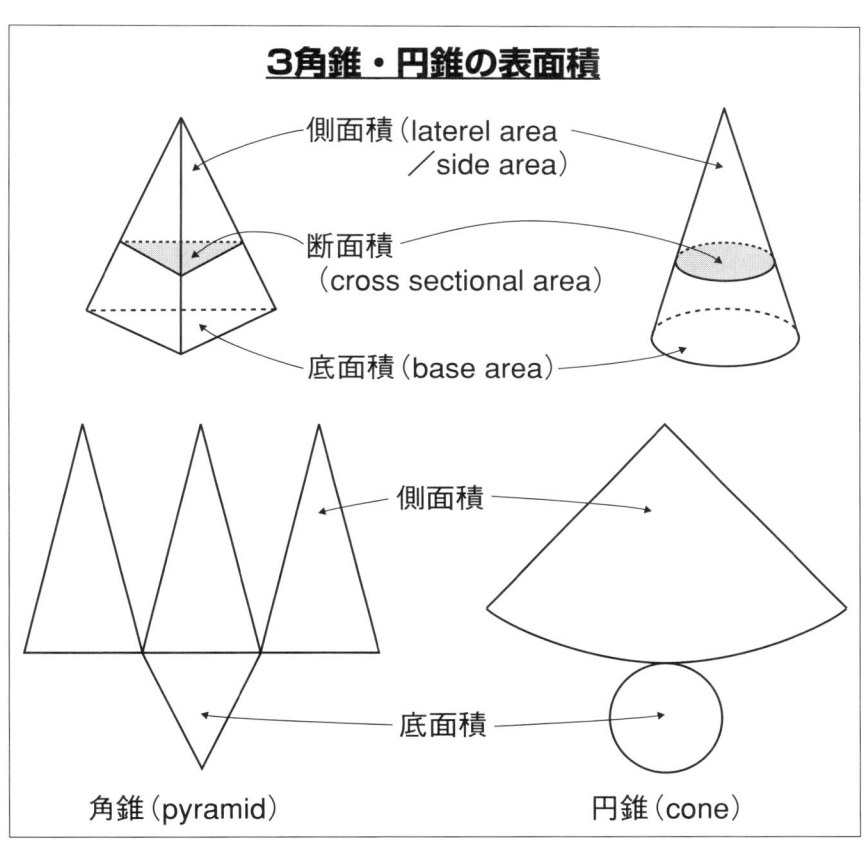

- 側面積（laterel area／side area）
- 断面積（cross sectional area）
- 底面積（base area）
- 側面積
- 底面積

角錐（pyramid）　　　円錐（cone）

球の表面積

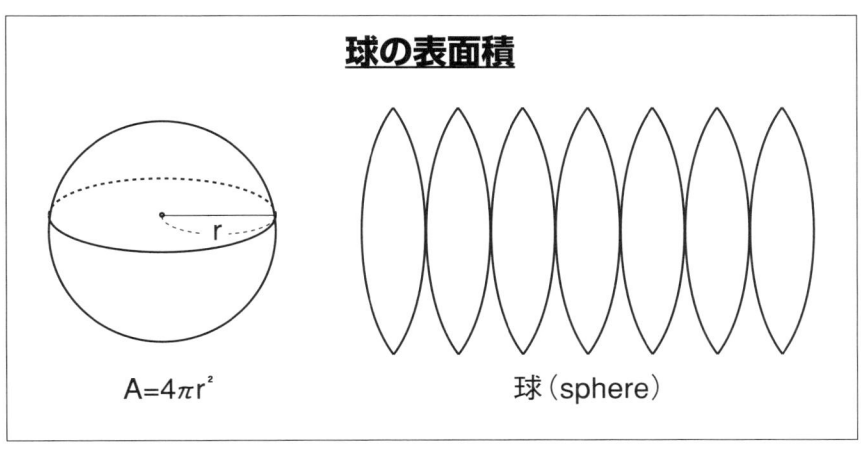

$A = 4\pi r^2$

球（sphere）

第5章●面積と体積／Areas and Volumes

●角柱と円柱の表面積は，底面積の2倍と側面積をたせば求められます。

The surface area of a prism or a cylinder can be found by adding twice the area of the base and the side area.

●角錐と円錐の表面積は，底面積と側面積をたせば，求められます。

The surface area of a pyramid or a cone can be found by adding the area of the base and the side area.

●1辺が15cmの立方体があります。表面積はいくらになりますか。

A certain cube has sides of 15 cm (fifteen centimeters). What is the cube's surface area?

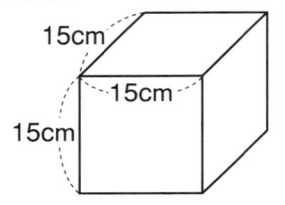

●底面が25cmと30cm，高さが45cmの直方体があります。表面積はいくらでしょうか。

Suppose a cuboid has a base of 25 cm (twenty-five centimeters) by 30 cm (thirty centimeters) and a height of 45 cm (forty-five centimeters). What is the cuboid's surface area?

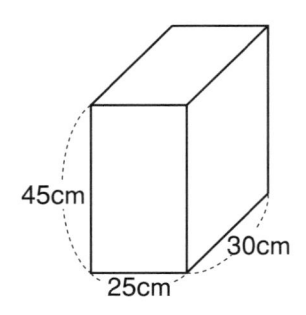

●底面の半径が45cm，高さが25cmの円柱の側面積と底面積をそれぞれ求めなさい。

Find the side area and the base area, respectively, of a cylinder whose radius of base is 45 cm (forty-five centimeters) and whose height is 25 cm (twenty-five centimeters).

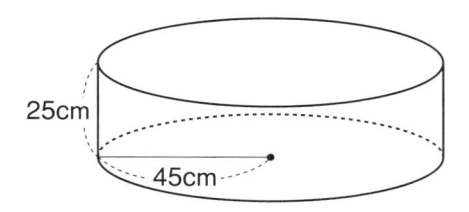

●つぎの展開図をもつ円錐の側面積・底面積・表面積はそれぞれいくらになりますか。円周率は3.14とします。

What is the side area, base area, and surface area, respectively, of a cone which has the following net? Let π be 3.14 (three point one four).

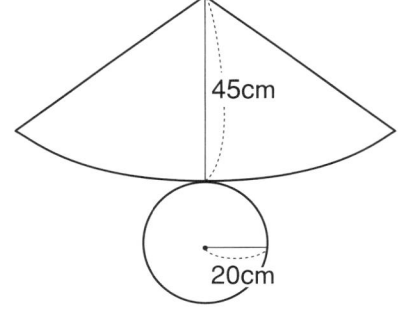

●半径 r の球の表面積は4π r × r です。

The surface area of a sphere of radius r is $4\pi r$ times r.

●球の表面積は大円の面積の4倍です。

The surface area of a sphere is four times the area of its great circle.

第5章●面積と体積／Areas and Volumes

④ 方体・柱・錐・球の体積

Volume of Cuboids, Prisms, Pyramids and Spheres

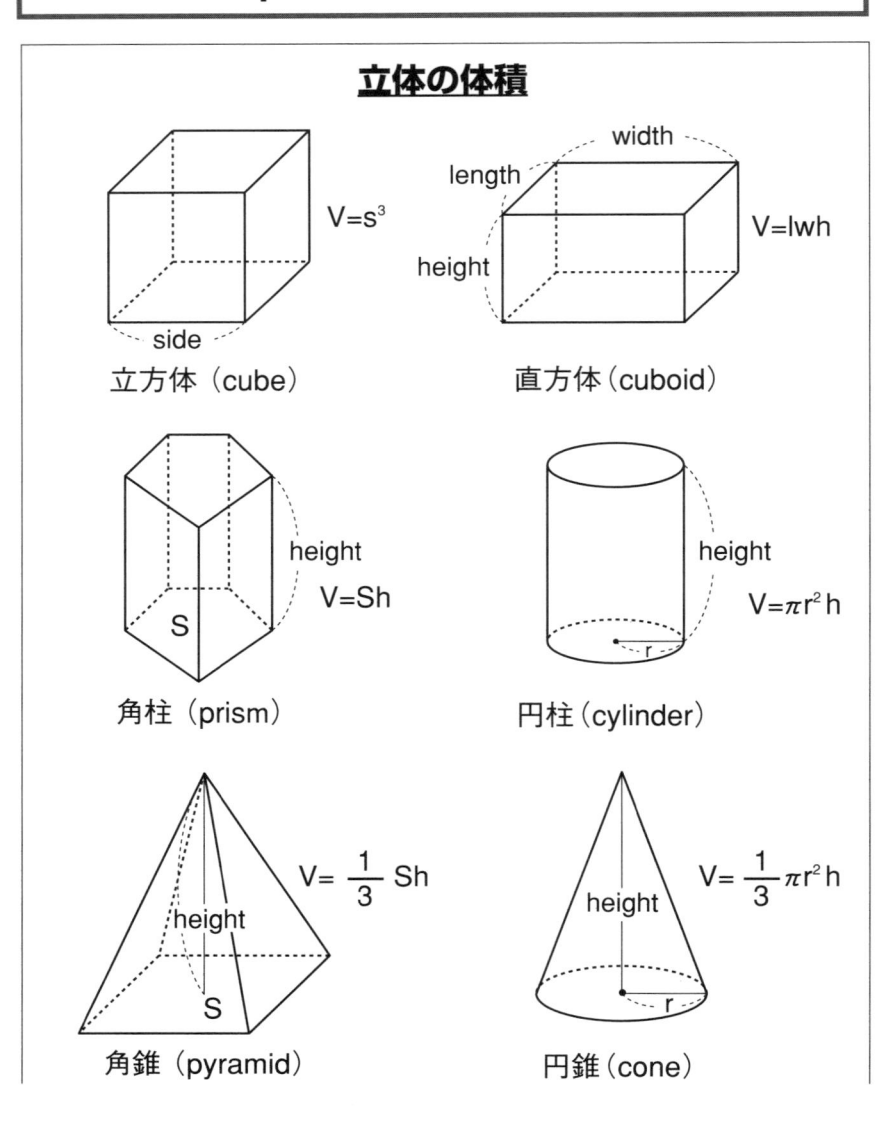

立体の体積

立方体 (cube) — $V = s^3$, side

直方体 (cuboid) — $V = lwh$, length, width, height

角柱 (prism) — $V = Sh$, height, S

円柱 (cylinder) — $V = \pi r^2 h$, height, r

角錐 (pyramid) — $V = \dfrac{1}{3} Sh$, height, S

円錐 (cone) — $V = \dfrac{1}{3} \pi r^2 h$, height, r

球(sphere)　　$V = \dfrac{4}{3}\pi r^3$

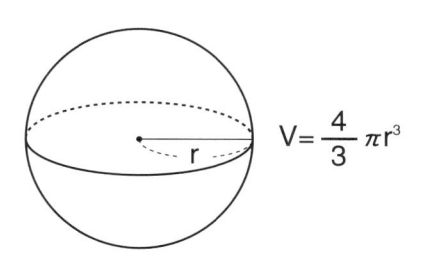

体積の単位の読み方

cm³（立方センチメートル）……………cubic centimeter
m³　（立方メートル）…………………cubic meter

●角柱も円柱も，その体積は底面積と高さの積で求められます。

The volume of either a prism or a cylinder can be obtained by the product of the area of the base and the height.

●角錐と円錐の体積は，底面積と高さが等しい角柱や円柱の体積の $\dfrac{1}{3}$ になります。

The volume of a pyramid or a cone is $\dfrac{1}{3}$ (one third) the volume of a prism or cylinder with the same base area and height.

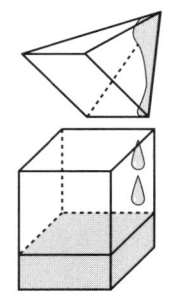

第5章●面積と体積／Areas and Volumes

●縦4m，横3m，深さ6mの大きな穴を埋めるのに，土はどのくらい必要ですか。

How much soil is needed to fill a big hole that is 4 m (four meters) long, 3 m (three meters) wide, and 6 m (six meters) depth?

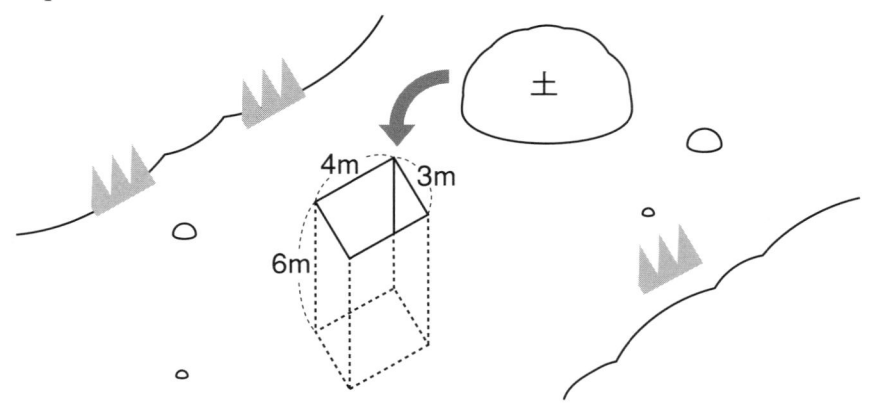

●底面の半径が3cm，高さが24cmの円柱の体積はいくらでしょうか。

What is the volume of a cylinder whose radius of base is 3 cm (three centimeters) and height is 24 cm (twenty-four centimeters)?

●つぎの展開図をもつ円柱の体積はいくらでしょうか。

What is the volume of a cylinder with the following net?

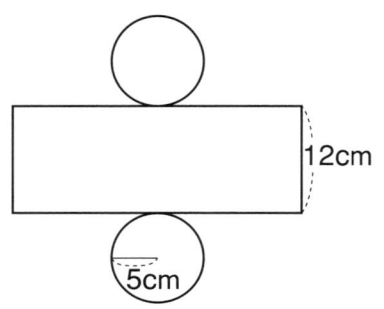

●つぎの図の角錐と円錐の体積を計算してください。

Calculate the volumes of the pyramid and the cone in the following figure.

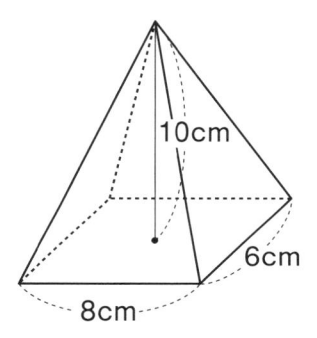

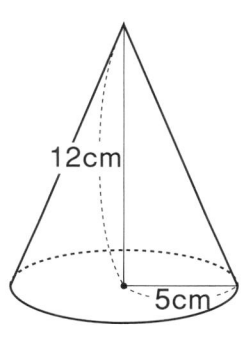

●内のりの半径が16cmの半球状をした容器の体積はいくらですか。円周率を3.14として求めてください。

What is the volume of a semispherical container whose internal radius is 16 cm (sixteen centimeters)? Let π be 3.14 (three point one four).

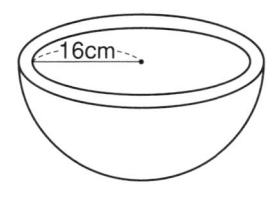

●体積は，1辺が1cmなり，1mなりの立方体が何個分あるかで表わします。

Volumes are expressed with how many cubes with side length 1 cm or 1 m are equivalent.

チャレンジ・テスト⑤ CHALLENGE 練習問題

① A square is made with a piece of string of 48 cm (forty-eight centimeters). What is the area of the square?
［48cmのヒモで正方形をつくりました。面積はいくらでしょうか。］

② Find the area of the land shown below.
［つぎの土地の面積を求めなさい。］

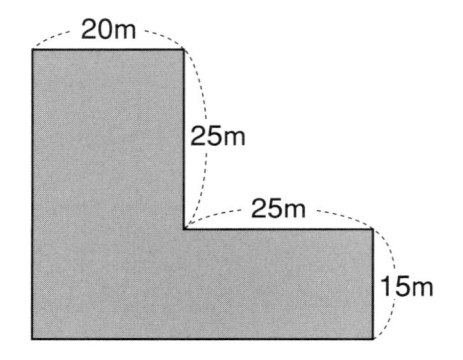

- 20m
- 25m
- 25m
- 15m

③ Find the volume of the shape shown below.
［つぎの体積を求めましょう。］

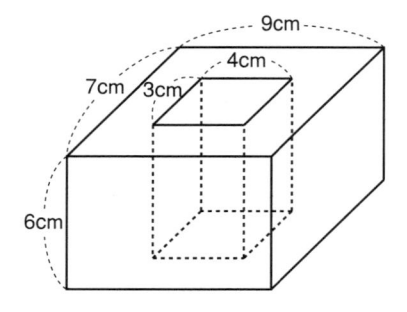

- 9cm
- 4cm
- 7cm
- 3cm
- 6cm

④ Using a board 2 cm (two centimeters) thick, a water tank was made whose outer size is 20cm (twenty centimeters) long, 24 cm (twenty-four centimeters) wide, and 17 cm (seventeen centimeters) high. Up to how many liters of water can be stored in the tank?
[厚さ2cmの板で，外回りの縦20cm，横24cm，高さ17cmの水槽を作りました。水は最大で何ℓはいりますか。]

⑤ Find the areas of the figures shown below.
[つぎの図形の面積を求めてください。]

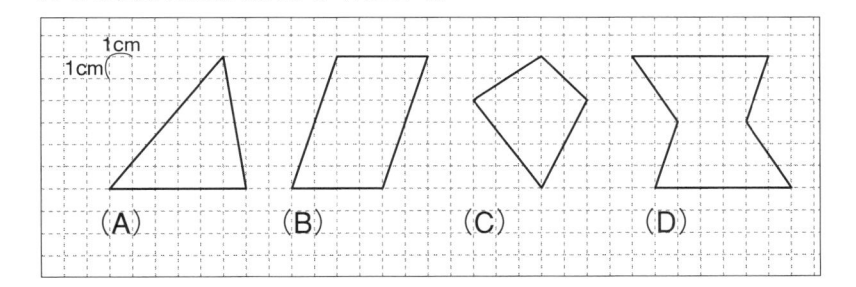

(A) (B) (C) (D)

⑥ What are the surface areas and the volumes of the figures shown below?
[つぎの図形の表面積と体積はいくらですか。]

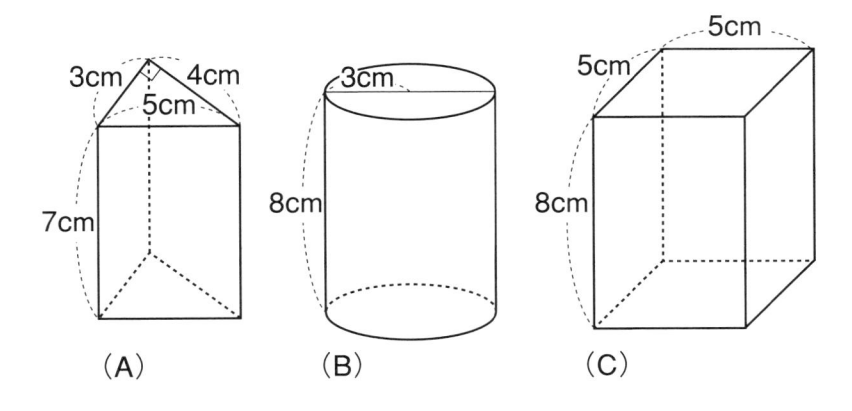

(A) (B) (C)

⑦ What is the area of a circle with a radius of 8 cm (eight centimeters)? Let $\pi = 3.14$ (pi equal three point one four).

[半径8cmの円の面積はいくらですか。π＝3.14として計算しましょう。]

⑧ Find the volume of the figure shown below. Use 3.14 (three point one four) for π (pi).

[つぎの図形の体積を求めましょう。円周率は3.14とします。]

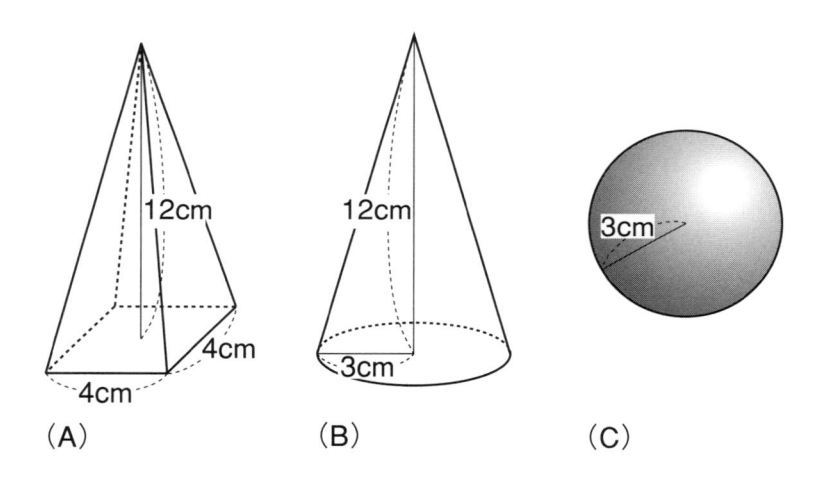

(A) 12cm, 4cm, 4cm

(B) 12cm, 3cm

(C) 3cm

第6章 表とグラフ

Tables and Graphs

① 表と棒グラフ・折れ線グラフ
Table, Bar Graph and Line Graph

表の名前

マジックの本数調べ ← 表題 (title)

色	黒	赤	青	緑	黄
本数(本)	4	8	5	6	7

↑項目 (item)

単位 (unit)

表 …… table

いろいろなグラフ

色えんぴつの本数調べ
(本)

棒グラフ (bar gragh／bar chart)

気温調べ

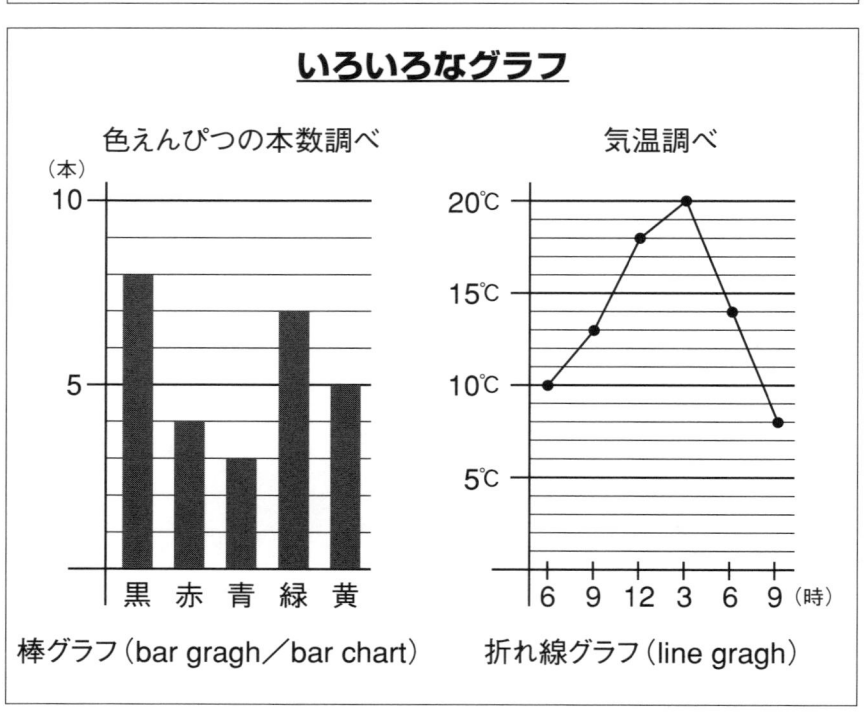

折れ線グラフ (line gragh)

グラフの名前

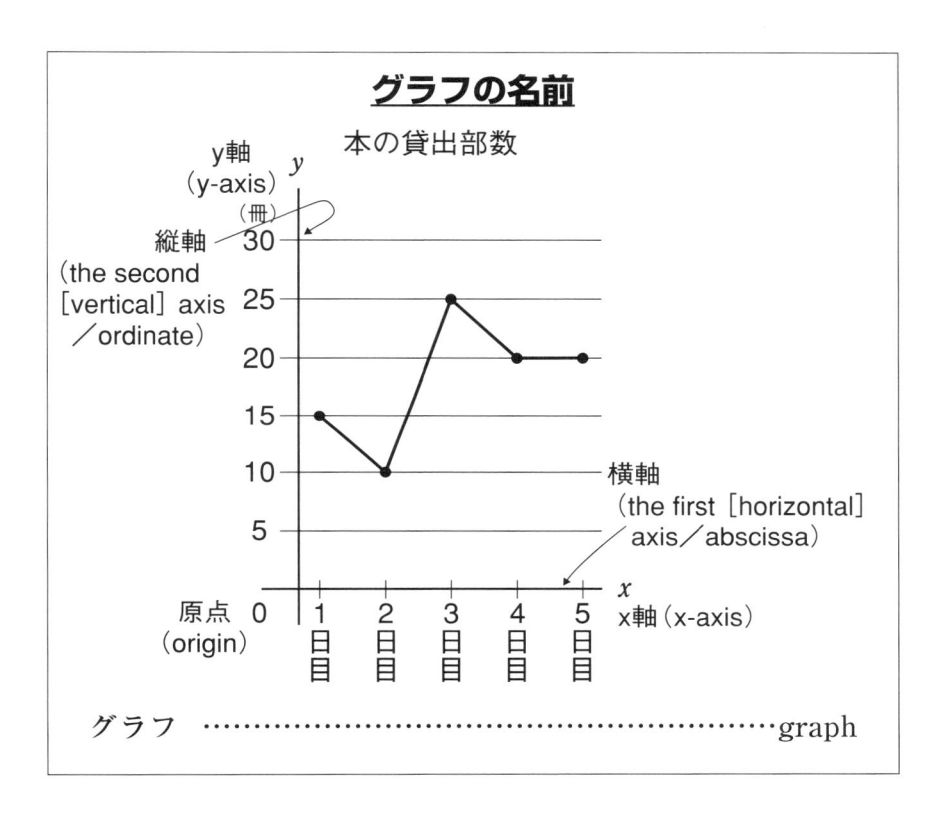

本の貸出部数

- y軸（y-axis）
- 縦軸（the second [vertical] axis／ordinate）
- 横軸（the first [horizontal] axis／abscissa）
- x軸（x-axis）
- 原点（origin）
- 1日目, 2日目, 3日目, 4日目, 5日目
- （冊）

グラフ ･････････････････････････････graph

●太郎さんは駐車場にある自動車の色を調べ，色別に表にまとめました。それを棒グラフにしてみましょう。

 Taro examined the colors of the cars in the parking lot and made a table with respect to color. Let's try to make it into a bar graph.

●気温や体重のように，あるものの変化の特徴をつかむには折れ線グラフが便利です。

 To show the characteristics of the change of something, such as the temperature or your weight, a line graph is useful.

❷ 百分率と円グラフ・帯グラフ
Percent, Circle Graph and Bar Graph

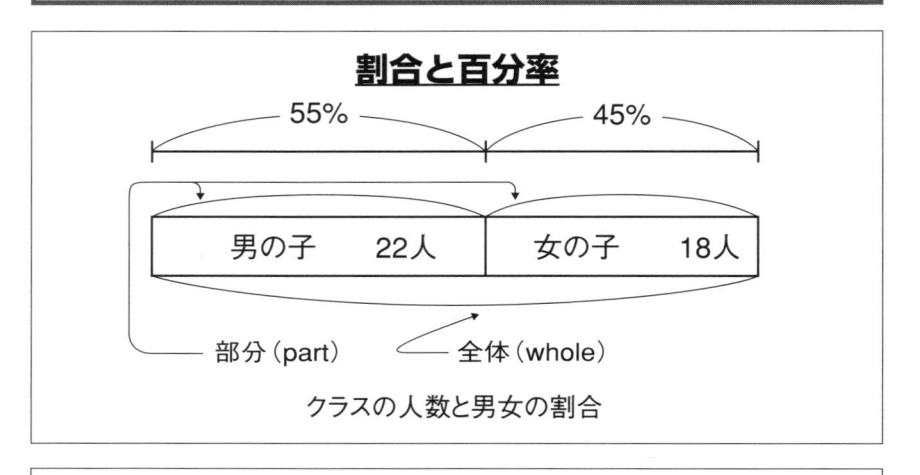

割合と百分率

```
       55%              45%
   ┌─────────┐     ┌─────────┐
   │ 男の子 22人 │ │ 女の子 18人 │
   └─────────┘     └─────────┘
      部分(part)      全体(whole)
```

クラスの人数と男女の割合

主要なことばと読み方

割合	rate
百分率	percent／percentage
1％	one percent
47％	forty-seven percent
34.5％	thirty-four point five percent
160％	one hundred [and] sixty percent
1割	one out of ten (a tenth, one tenth)
1分	one percent (a hundredth, one hundredth)
1厘	a (one) thousandth
4割7分	forty-seven percent (hundredths)
3割4分5厘	three hundred [and] forty-five thousandths
16割	sixteen tenths

円グラフと帯グラフ

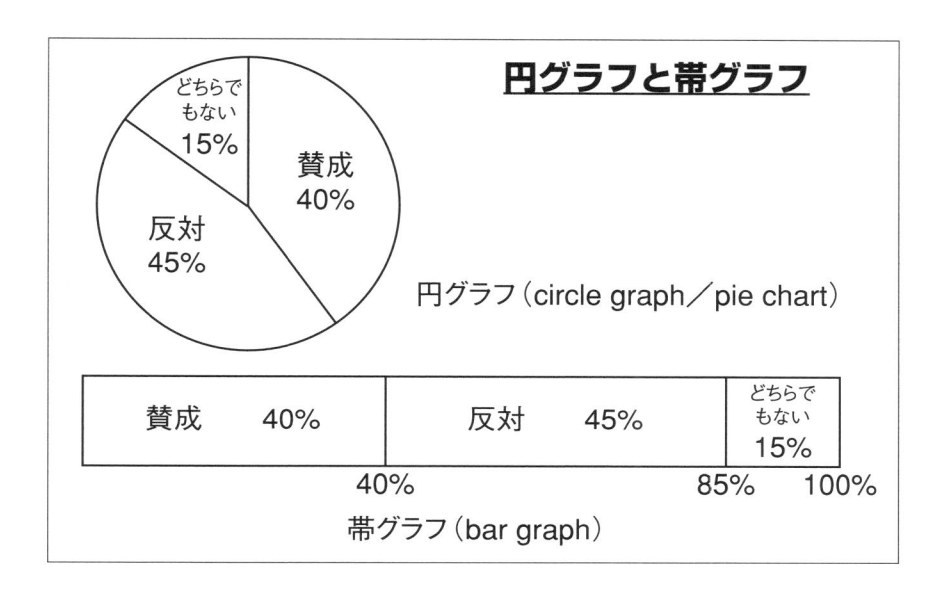

円グラフ（circle graph／pie chart）

- 賛成 40%
- 反対 45%
- どちらでもない 15%

帯グラフ（bar graph）

●百分率とは，元にする量を100としたときの割合のことで，0.01を1パーセントといい，1％と書きます。

A percentage is the rate of something with the base quantity taken as 100 (one hundred). 0.01 (One point zero one) is called 1 (one) percent and written as 1%.

●定価350円の文具が2割引きで売られています。いくらで買えますか。

Some stationery priced at 350 (three hundred [and] fifty) yen is sold at a 20 (twenty) percent discount. How much does it cost?

●80人にアンケート用紙を配ったところ，56人が回答をしてくれました。回答率は何パーセントでしょう。

When a questionnaire was distributed to 80 (eighty) people, 56 (fifty-six) responded to it. Show the response rate as a percentage.

③ 平均・分布・延べと柱状グラフ
Average, Distribution, Total and Histogram

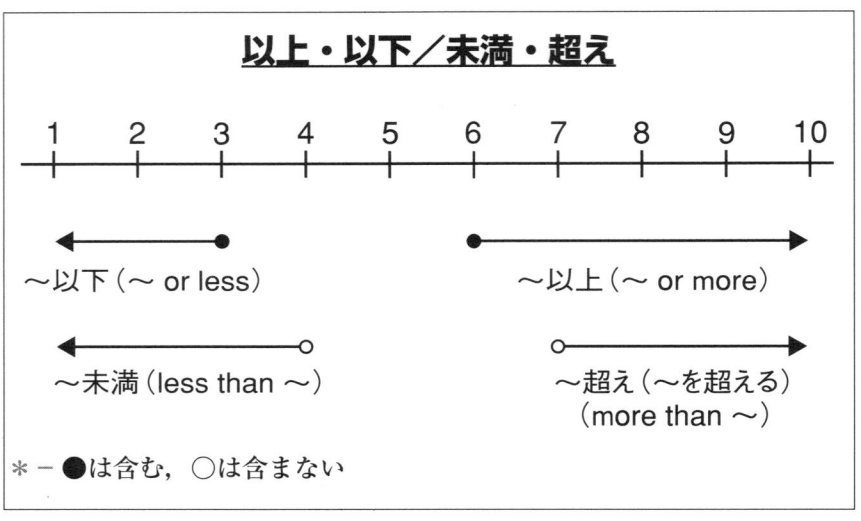

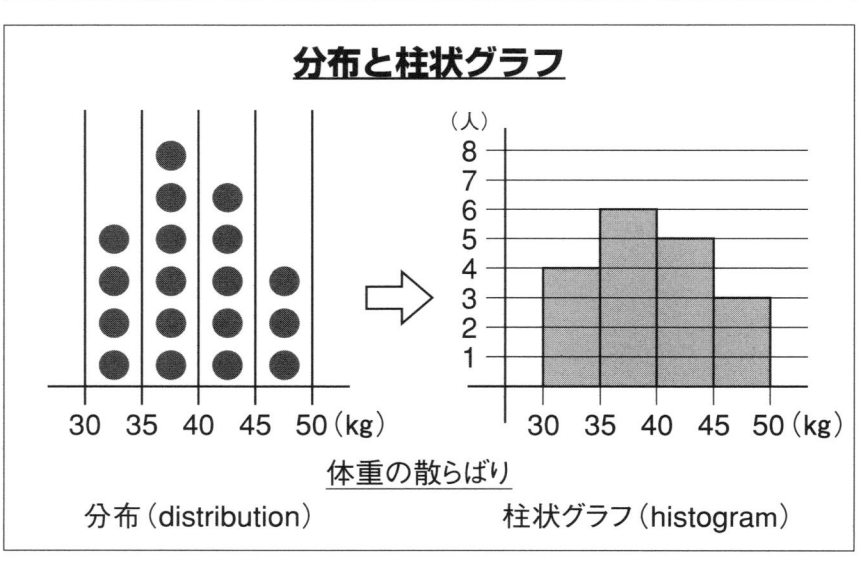

以上・以下／未満・超え

～以下（～ or less）

～以上（～ or more）

～未満（less than ～）

～超え（～を超える）（more than ～）

＊ －●は含む，○は含まない

分布と柱状グラフ

体重の散らばり

分布（distribution）　　　柱状グラフ（histogram）

主要なことば

平均	average
分布（散らばり）	distribution (dispersion)
延べ	total (gross, cumulative)
端を含んで	inclusive
端を含まないで	exclusive

●柱状グラフは棒グラフと似ていますが，棒と棒の間をつけます。

A histogram closely resembles a bar graph but in a histogram there are no spaces between the bars.

●表やグラフは目的にあわせて使いわけるのがコツです。

The point is to choose a type of tables and graphs which is appropriate to the purpose.

●この部屋には身長が150cm以上で175cm以下の人が6人います。

In this room there are six people whose height is between 150 cm (one hundred [and] fifty centimeters) and 175 cm (one hundred [and] seventy-five centimeters) inclusive.

●12歳未満の人は入館料が500円です。

For those under the age of 12 (twelve), the entrance fee is 500 (five hundred) yen.

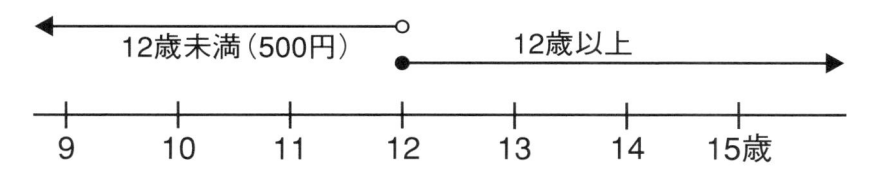

●時速80kmを超えると，ルール違反になります。

Speed exceeding 80 km (eighty kilometers) per hour are in violation of the law.

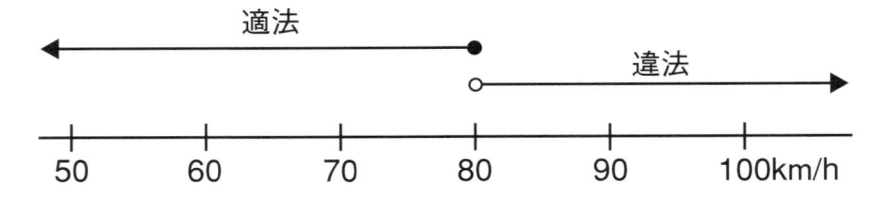

適法　違法

50　60　70　80　90　100km/h

●遅刻者が1日目は4人，2日目は5人，3日目は2人，4日目は0人，5日目は8人でした。延べ何人でしょう。

The number of latecomers was 4 (four) on the 1st (first) day, 5 (five) on the 2nd (second), 2 (two) on the 3rd (third), 0 (zero) on the 4th (fourth), and 8 (eight) on the 5th (fifth) day. How many in total?

●一郎さんがボール投げを5回したら，18m，21m，22m，15m，12mでした。平均何mになりますか。また，平均からの散らばりの範囲はいくらですか。

Ichiro threw a ball 5 (five) times for distances of 18 m (eighteen meters), 21 m (twenty-one meters), 22 m (twenty-two meters), 15 m (fifteen meters), and 12 m (twelve meters). How many meters was his average throw? Also, what is the range of dispersion from the average for his throws?

1回目	2回目	3回目	4回目	5回目
18m	21m	22m	15m	12m

チャレンジ・テスト ⑥ CHALLENGE 練習問題

① Examining the numbers of students who borrowed any book from the library last week with respect to school year resulted as follows. Let's make a table and try to make it into a bar graph. 1st (First) year: 14 (fourteen) students, 2nd (second) year: 23 (twenty-three) students, 3rd (third) year: 18 (eighteen) students, 4th (fourth) year: 25 (twenty-five) students, 5th (fifth) year: 17 (seventeen) students, 6th (sixth) year: 20 (twenty) students.

〔先週,図書室から本を借りた人数を学年ごとに調べたら,つぎのようになりました。それを表にまとめ,棒グラフにしてみましょう。1年－14人,2年－23人,3年－18人,4年－25人,5年－17人,6年－20人〕

② Regularly examining the temperature for one day resulted in the following. Let's make it into a line graph. 6 (six) a.m.: 6℃ (six degrees Celsius), 8 (eight) a.m.: 12℃ (twelve degrees Celsius), 10 (ten) a.m.: 16℃ (sixteen degrees Celsius), 12 (twelve) noon: 19℃ (nineteen degrees Celsius), 2 (two) p.m.: 22℃ (twenty-two degrees Celsius), 4 (four) p.m.: 15℃ (fifteen degrees Celsius), 6 (six) p.m.: 13℃ (thirteen degrees Celsius).

〔1日の気温を調べたら,つぎのようになりました。折れ線グラフにしてみましょう。午前6時－6℃,8時－12℃,10時－16℃,12時－19℃,午後2時－22℃,4時－15℃,6時－13℃〕

③ 80 (Eighty) children have gathered together. Among them, 48 (forty-eight) children are girls. What percent are boys?
［80人の子どもたちが集まっています。そのうち48人は女の子です。男の子は何パーセントでしょうか。］

④ Here are ribbons of various colors. Calculate the percentage of each and then make a circle graph and a bar graph. Red: 24 (twenty-four) ribbons, blue: 30 (thirty), purple: 12 (twelve), yellow: 48 (forty-eight), green: 36 (thirty six), and the total: 150 (one hundred and fifty).
［ここにいろいろな色のリボンがあります。それぞれの割合を調べて円グラフと帯グラフにしましょう。赤－24本，青－30本，紫－12本，黄色－48本，緑－36本，合計－150本］

⑤ The numbers of students who forgot to bring something with them last week were as follows. How many students forgot something per day on average? Also, what is the total number of the students who forgot something? Monday: 5 (five), Tuesday: 10 (ten), Wednesday: 0 (zero), Thursday: 7 (seven), Friday: 13 (thirteen), and Saturday: 1 (one).
［先週，忘れ物をした人数はつぎのようでした。1日平均，何人が忘れ物をしましたか。また，延べ何人になりますか。月曜日－5人，火曜日－10人，水曜日－0人，木曜日－7人，金曜日－13人，土曜日－1人。］

⑥ The results of the test are as follows. Let's express their dispersion in a histogram. 90 (ninety) points and over: 2 (two), from 80 (eighty) (inclusive) to 90 (ninety) (exclusive): 5 (five), from 70 (seventy) (inclusive) to 80 (eighty) (exclusive): 7 (seven), from 60 (sixty) (inclusive) to 70 (seventy) (exclusive): 4 (four), from 50 (fifty) (inclusive) to 60 (sixty) (exclusive): 2 (two), and total: 20 (twenty).

［テストの結果はつぎの通りです。散らばりぐあいを柱状グラフにしましょう。90点以上－2人，90点未満80点以上－5人，80点未満70点以上－7人，70点未満60点以上－4人，60点未満50点以上－2人，合計－20人。］

Tea Room ❸

書籍・新聞・雑誌

① －○ページ ──────── Page 35（35ページ）
　　　Turn to page 32.（32ページを開けなさい）
　　　The list of contents is on page 8.
　　　（目次は8ページ目にあります）
② －第○欄 ──────── col.5（第5欄）
　　　See col.2, p.12.（12ページの第2欄を見てください）
③ －第○巻第○号 ──────── Vol.10, No.6（第10巻第6号）
　　　This magazine is Vol.5, No.3.
　　　（この雑誌は第5巻第3号です）
④ －部数 ──────── 3000部（3000 copies）
　　　Average daily sales : 2,500,000 copies.
　　　（1日の平均販売部数：250万部）
⑤ －○月○日号 ──────── the issue for March 5
　　　　　　　　　　　　　　＝ the March 5 issue（3月5日号）
　　　Is that the issue for July 7？
　　　（あれは7月7日号ですか）
⑥ －第○章 ──────── the 4th chapter ＝ Chapter 4（第4章）
　　　Chapter 1 is a general introduction.
　　　（第1章は概論です）
⑦ －第○章〜第□章 ──────── Chapters 3-6（第3章から第6章）
　　　Chapters 2-5 are about report.
　　　（第2章から第5章までは報告です）
⑧ －著作権 ──────── Copyright ©2005（2005年著作権取得）

# 第7章
比・比例と場合の数

Ratios, Proportion and Number of Cases

① 比と比例
Ratios and Proportion

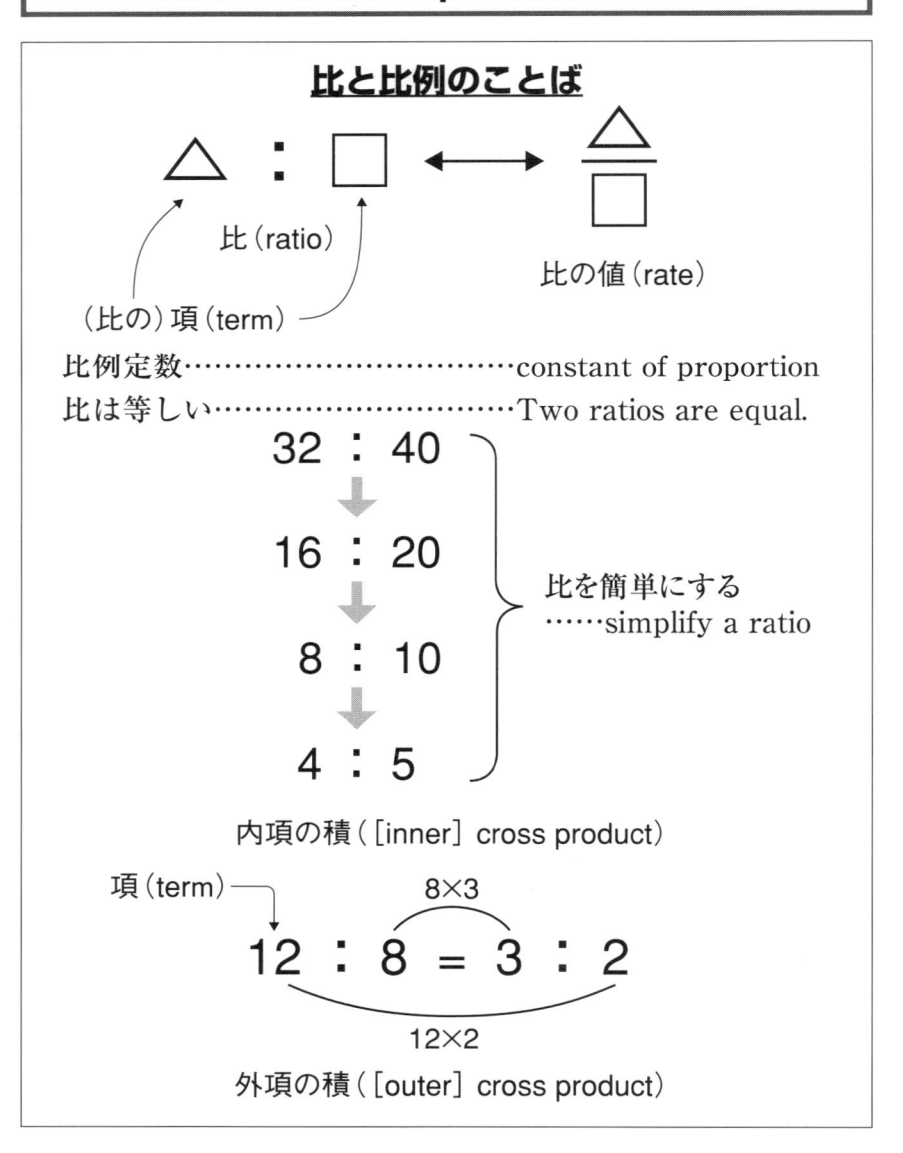

比と比例のことば

△ : □ ↔ △/□

比 (ratio)
(比の) 項 (term)
比の値 (rate)

比例定数……………………………constant of proportion
比は等しい…………………………Two ratios are equal.

32 : 40
↓
16 : 20
↓
8 : 10
↓
4 : 5

比を簡単にする……simplify a ratio

内項の積（[inner] cross product）

項 (term)

8×3
12 : 8 = 3 : 2
12×2

外項の積（[outer] cross product）

正比例

水の量 (dℓ)	1	2	3	4	5	6	…	…
水の深さ (cm)	3	6	9	12	15	18	…	…

2倍　3倍

1 : 3 = 2 : 6
　　 = 3 : 9 = 4 : 12
比例式（proportion）

（正）比例 …………………………………… [direct] proportion

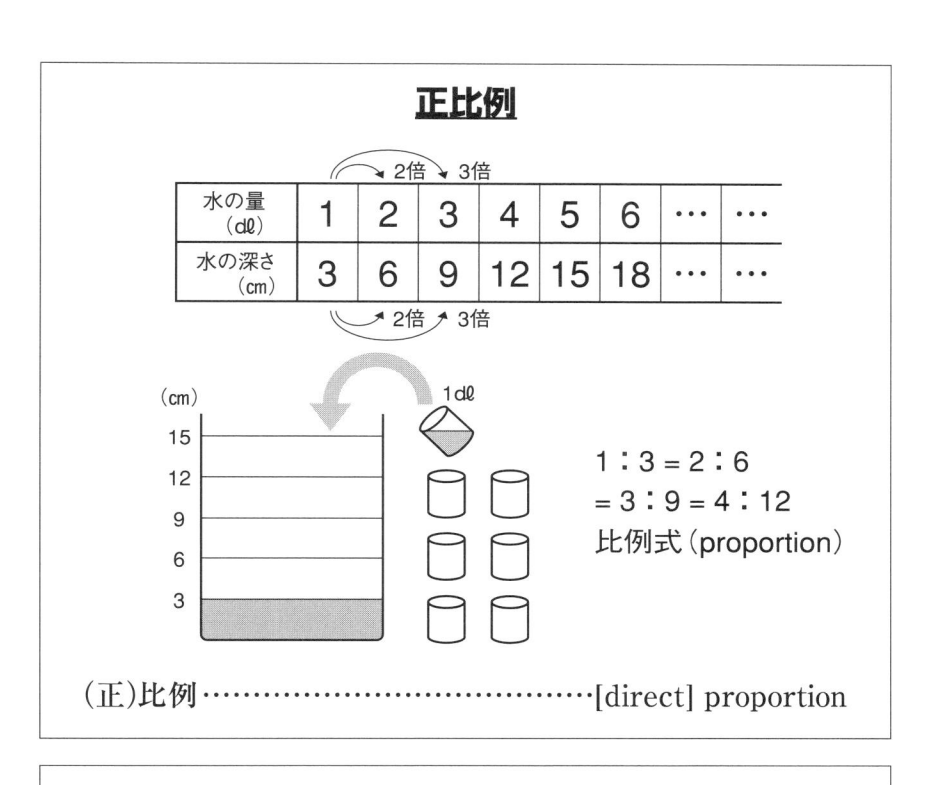

反比例

面積12cm²の長方形の縦と横の長さ

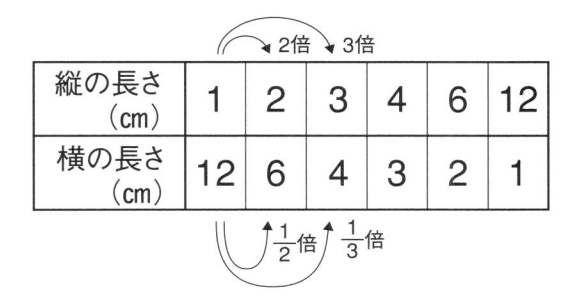

縦の長さ (cm)	1	2	3	4	6	12
横の長さ (cm)	12	6	4	3	2	1

2倍　3倍

$\frac{1}{2}$倍　$\frac{1}{3}$倍

反比例 …………………………………… inverse proportion

第7章●比・比例と場合の数／Ratios, Proportion and Number of Cases

比例の表わし方

○が□に比例する ……… ○ is [directly] proportional to □
○が□に反比例する …… ○ is inversely proportional to □
2 : 3 ………………… two to three
A : B ………………… A to B

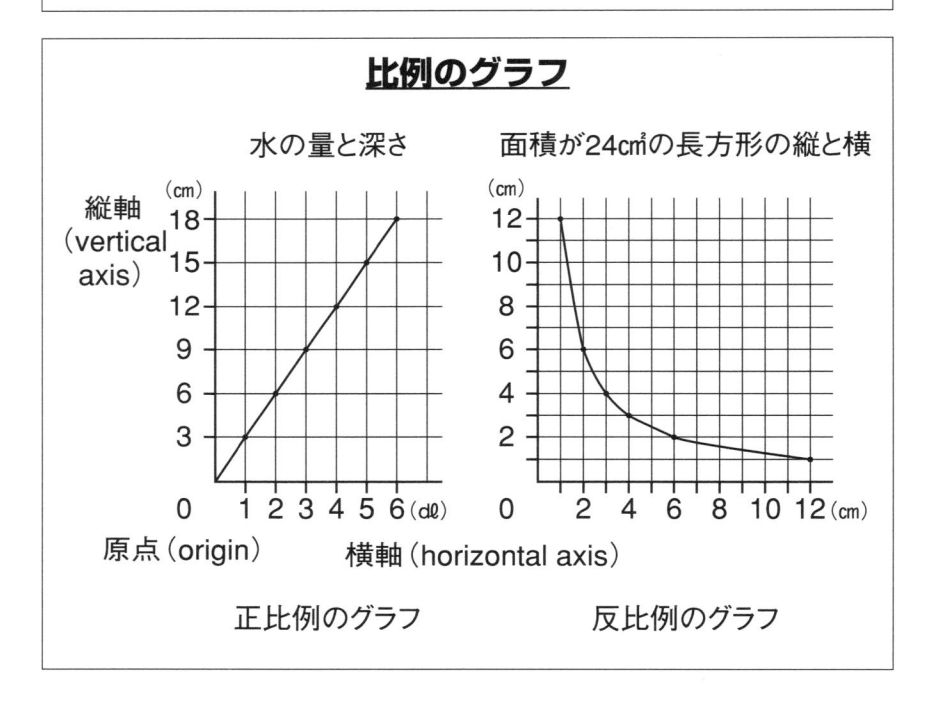

比例のグラフ

水の量と深さ

縦軸 (vertical axis)

原点 (origin)　　横軸 (horizontal axis)

正比例のグラフ

面積が24cm²の長方形の縦と横

反比例のグラフ

● 比はできるだけ小さい整数の比に直して表わします。
A ratio is expressed as a ratio of integers as small as possible.

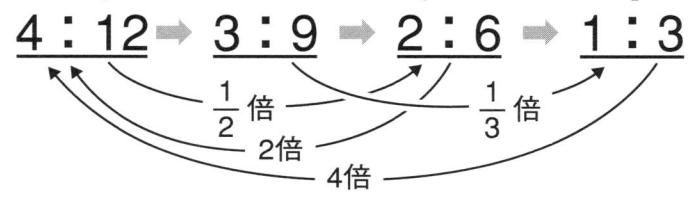

$$4:12 \Rightarrow 3:9 \Rightarrow 2:6 \Rightarrow 1:3$$

$\frac{1}{2}$倍　　$\frac{1}{3}$倍
2倍
4倍

① ―比と比例／Ratios and Proportion

●比は同じ数をかけても，同じ数でわっても等しくなります。

A ratio remains the same even when multiplied by the same number, or when divided by the same number.

$$2:3 = 4:6 = 8:12$$

2でわる　　2をかける

●正比例のグラフは右上がりの直線になります。

A graph of [direct] proportion becomes an ascending (upward-sloping) straight line.

●反比例のグラフは曲線になります。

A graph of inverse proportion becomes a curve.

●赤組と白組の人数の比は3対2です。

The ratio of the number of red team members to white team members is 3 (three) to 2 (two).

●自動車の速度と進む距離は比例します。グラフで表わしましょう。

The speed of a car is [directly] proportional to the distance it runs. Show this as a graph.

●長方形の面積が一定のとき，縦の長さと横の長さは反比例します。

When the area of a rectangle is constant, the length is inversely proportional to the width.

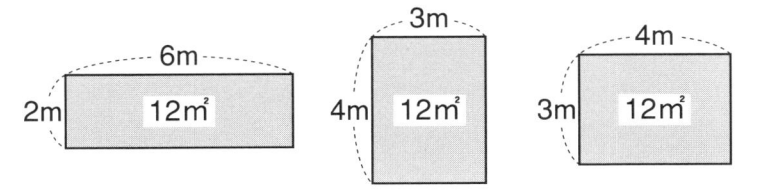

第7章●比・比例と場合の数／Ratios, Proportion and Number of Cases

❷ 拡大と縮小（相似）
Enlargement and Reduction(Similarity)

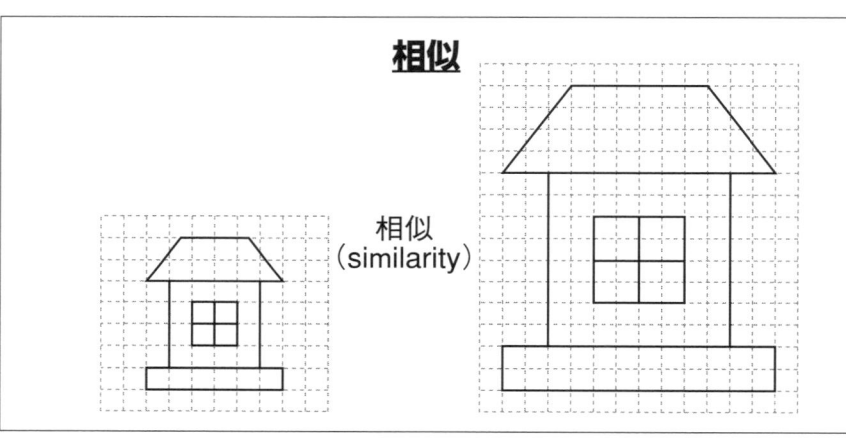

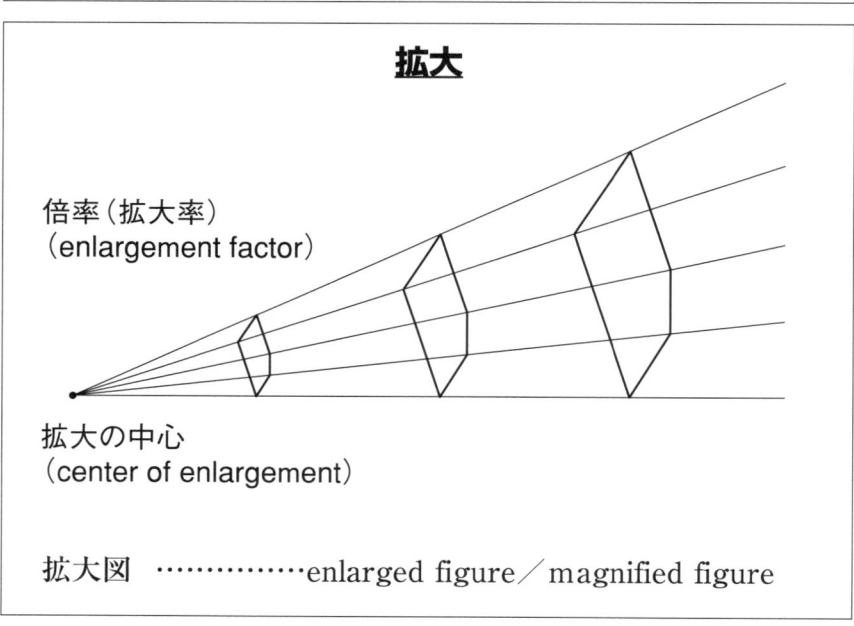

相似
相似（similarity）

拡大
倍率（拡大率）
(enlargement factor)

拡大の中心
(center of enlargement)

拡大図 ………enlarged figure／magnified figure

縮小

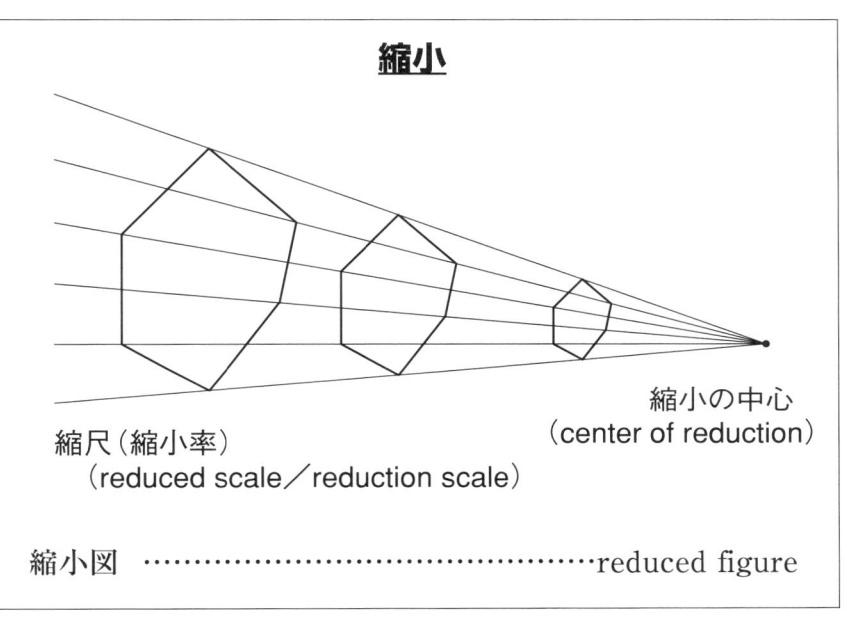

縮尺（縮小率）
(reduced scale／reduction scale)

縮小の中心
(center of reduction)

縮小図 ································· reduced figure

相似な3角形

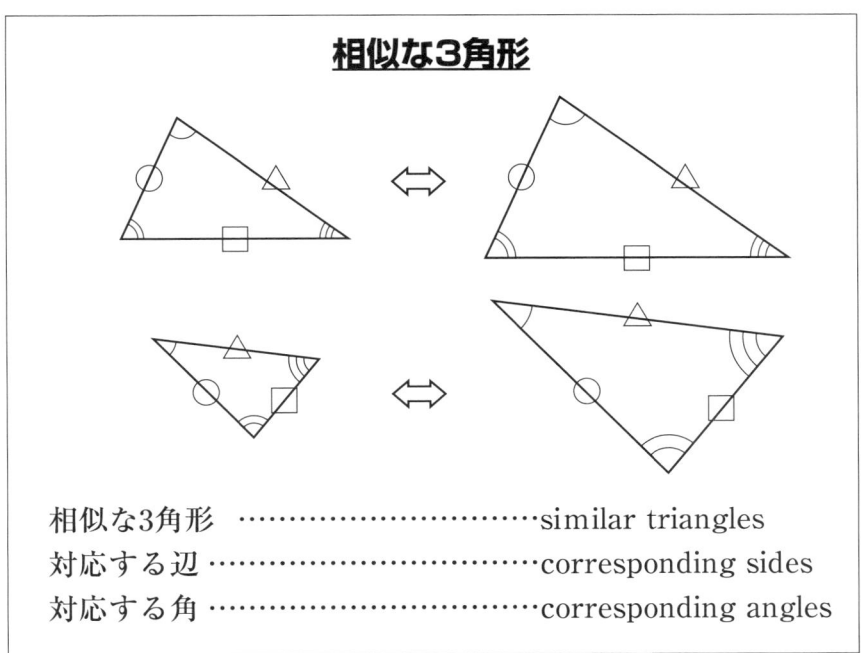

相似な3角形 ···························· similar triangles
対応する辺 ······························ corresponding sides
対応する角 ······························ corresponding angles

第7章●比・比例と場合の数／Ratios, Proportion and Number of Cases

相似な多角形

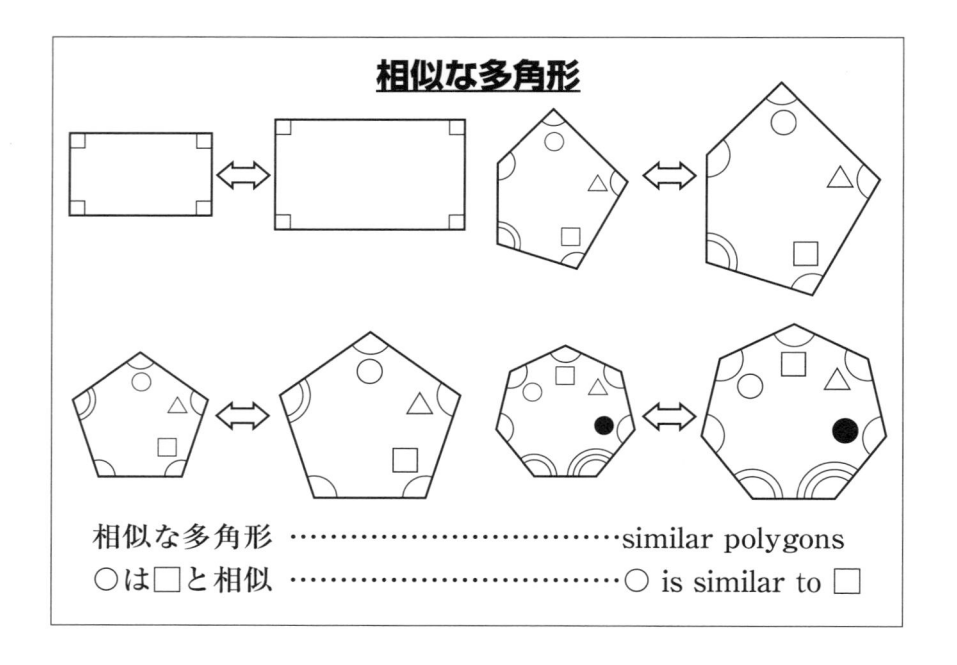

相似な多角形 ……………………………… similar polygons
○は□と相似 ……………………………… ○ is similar to □

●角度は変えずに辺の長さだけを同じ倍率で大きくしたものを拡大図といいます。

A figure where all sides are magnified by the same enlargement factor without changing any angles is called an enlarged (a magnified) figure.

●角度は変えずに辺の長さだけを同じ倍率で小さくしたものを縮小図といいます。

A figure where all sides are reduced by the same reduction scale without changing any angles is called a reduced figure.

●相似な多角形は対応する角が等しく，対応する辺の比がすべて同じです。

In similar polygons, their corresponding angles are equal and the ratios of their corresponding sides are all equal.

●2つの3角形の対応する2つの角が等しければ，相似です。

Two triangles are similar if their two corresponding angles are equal.

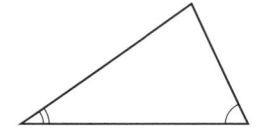

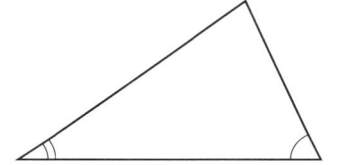

●拡大図でも縮小図でも，角度は変わりません。辺だけが，拡大図では倍率に応じて長くなり，縮小図では縮尺に応じて短くなります。

Any angle does not change in either enlarged or reduced figures. Only sides are lengthened in enlarged figures according to the enlargement factors, and are shortened in reduced figures according to the reduction scales.

●面積は，拡大図では倍率×倍率，縮小図では縮尺×縮尺になります。

Areas are multiplied by the enlargement factor squared in enlarged figures, and by the reduction scale squared in reduced figures.

③ 場合の数（順列・組合せ・確率）

Permutations, Combinations and Probability

いろいろなタイプ

▶ Aタイプ ── いくつものなかから2つを組合せる

	A	B	C	D	E
A	╲	○	○	●	○
B	●	╲	●	○	●
C	●	○	╲	○	○
D	○	●	●	╲	●
E	●	○	●	○	╲

例 ── 5チームによるリーグ戦の対戦成績

▶ Bタイプ ── いくつかのものの並べ方

```
     ┌ B ── C
A ──┤
     └ C ── B

     ┌ A ── C
B ──┤
     └ C ── A

     ┌ A ── B
C ──┤
     └ B ── A
```

例 ── 3人で走るリレーの順番

▶Cタイプ ── いくつかのものの重複を許した並べ方

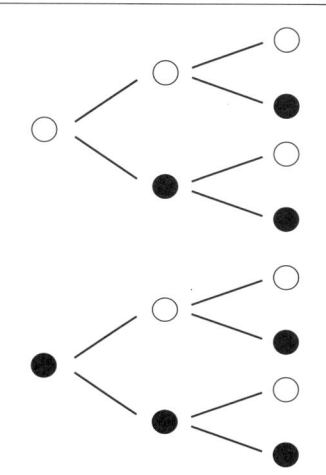

例 ── コインを3回投げたときの表と裏の出方

主要なことば

場合の数 ……………………number of events (outcomes)
並べ方 ………………………arrangement
順列 …………………………permutation
組合せ ………………………combination
樹形図 ………………………tree diagram
起こりやすさ（確からしさ）……likelihood
確率 …………………………probability
（コインの）表／裏………head (obverse)／tail (reverse)
$n \times (n-1) \times (n-2) \times (n-3) \cdots$
    …………n times [opening bracket] n minus 1 (one) [closing bracket] times [opening bracket] n minus 2 (two) [closing bracket] times [opening bracket] n minus 3 (three) [closing bracket], and so on

●10円玉を2回投げたとき，表と裏の出方には何通りあるでしょうか。

When tossing a 10 (ten) yen coin twice, how many permutations of heads and tails are possible?

●赤・青・黄・緑のカードが1枚ずつ全部で4枚あります。この中から2枚を選んで組をつくります。全部で何通りできるでしょうか。

There are four cards: red, blue, yellow, and green. When selecting a pair of cards from among them, how many outcomes are possible?

	赤	青	黄	緑
赤		①	②	③
青			④	⑤
黄				⑥
緑				

●1組から5組まで5クラスで野球の試合をします。対戦の組合せは何通りあるでしょうか。

Five classes, from class 1 (one) to class 5 (five), are going to play baseball games. How many combinations of matches are possible?

チャレンジ・テスト ⑦ CHALLENGE 練習問題

① Simplify the following ratios: ① 63 : 56 (sixty-three to fifty-six), ② $\frac{3}{5} : \frac{4}{5}$ (three fifths to four fifths), ③ $\frac{3}{4} : \frac{5}{8}$ (three over four to five over eight).

［つぎの比を簡単にしましょう。
① 63 : 56　② $\frac{3}{5} : \frac{4}{5}$　③ $\frac{3}{4} : \frac{5}{8}$ ］

② To make soup, soup concentrate and hot water are mixed together at the ratio of 2 : 5 (two to five). How many milliliters of hot water is needed when there is 50 $m\ell$ (fifty milliliters) of concentrate?

［濃縮スープとお湯を2対5の比でまぜてスープをつくります。濃縮スープ50 $m\ell$ のとき，お湯は何 $m\ell$ 必要でしょうか。］

③ How long is Taro's shadow?

［太郎君の影の長さはいくらでしょうか。］

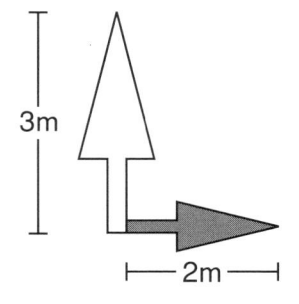

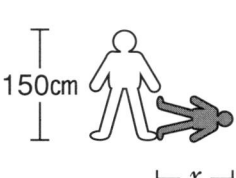

④ If the triangle shown below is enlarged by a factor of 3 (three), what will be the area of the enlarged figure?

Also, if the triangle is reduced by $\frac{1}{2}$ (one half), what will be the area of the reduced figure?

［つぎの3角形の，3倍の拡大図の面積はいくらですか。また，半分の縮小図の面積はいくらですか。］

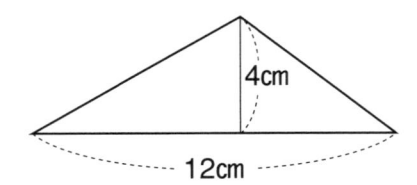

⑤ The figure below shows the vegetable garden of my house with a reduction scale of $\frac{1}{200}$ (one over two hundred). Find its actual area.

［下図はわが家の菜園を $\frac{1}{200}$ で表わした図です。実際の面積を求めてください。］

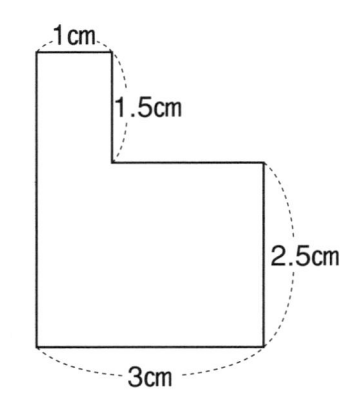

⑥ How many times as large as the original figure is the area of a triply magnified figure of a rectangle?

［ある長方形を3倍に拡大したとき，面積は元の図形の何倍になりますか。］

⑦ Matches will be held among 5 (five) classes. Each class will meet each of the other classes only once. How many combinations of matches will there be?
［5クラスで試合をします。どのクラスとも1回ずつ対戦するとき，対戦の組合せは何通りありますか。］

⑧ There are 3 (three) cards: A, B and C. How many ways are there to arrange them in a row?
［A，B，Cの3枚のカードがあります。並べる順は何通りありますか。］

⑨ When tossing a coin three times, how many permutations of heads and tails are possible?
［コインを3回投げたとき，表と裏の出方には何通りありますか。］

Tea Room ❹

通貨

① − 通貨単位の記号（＄，￥など）と数字で表わします。記号と数字の間には通常はスペースを入れません。
　　＄700（700ドル），￥3,500,000（350万円），
　　€5,600（5,600ユーロ）

② − 1ドル以上のときはドル記号を使い，1ドル未満のときはドル記号かセント記号を使います。イギリスの通貨であるポンド（£）とペンス（p）も同様です。
　　＄34.64（34.64ドル＝34ドル64セント），
　　＄.68＝68¢（68セント）

③ − 100万以上の切りの良い金額はmillionやbillionなどの単語を使って表わします。
　　＄458 million（4億5800万ドル），
　　＄3.547 billion（35億4700万ドル）
　　また，その場合、millionをM，billionをBと省略し，数字との間にスペースを入れずに表わすことがあります。
　　＄10M（1000万ドル），＄3B（30億ドル）

④ − 範囲を示す場合はどちらの数字にも単位記号をつけます。
　　＄650 − ＄850（650ドルから850ドル）

⑤ − 通貨単位を記号と単語で二重に表記しないように気をつけましょう。
　　￥652 yen（×）

●解説——数の英語表現，ここがポイント

英語による数の表現や表記にはルールがあります。とくに基数の読み方は算数・数学の基礎になります。ポイントになる，おもな原則をまとめておきます。

◆数の表記

① 1から20までの数はスペルアウトすることが多く，とくに文頭では極力スペルアウトします。
  Five boys are in school. (○)
  5 boys are in school. (×)

② 数字つきの名詞（例.「30冊の本」）を形容詞的に用いる場合は数字のあとにハイフンをいれます。
  a 30-book box（30冊入りの箱）
  a 120-minute tape （120分テープ）
  ＊このとき，数字がつく名詞には複数形の -s はつけません。

③ 数自体を複数形にする場合には，-s をつけます。
  He won the hand with a pair of fives.
  （彼は5のワン・ペアでその勝負に勝った）
  He looked like he was in his 30s.
  （彼は30歳代のように見えた）

④ 以下の場合は通常，数字で表わします。
  正確な数量（5 m, 4.8 g, 3 in.）／正確な金額（＄52,

¥3456）／日付（May 23, 2006）／ページ番号（page 78）／パーセント（48%）／番地（562 Theresa Street）／道路番号（Interstate 83）

⑤ 以下の場合は桁を表わすカンマをつけません。
　　　電話番号（03-3456-3355）／通し番号（56784325）／番地（1675 North Street）／小数点以下の数（3.6724）／西暦年（2007, July 25）

⑥ 数の範囲を表わすときはハイフンを使います。
　　　詳記／253-367 または from 253 to 367
　　　略記／・範囲を表わす数字の場合，後者の数字を略記することがあります。
　　　　　　（1234-1275──→1234-75）
　　　　　・西暦年（1946-1978──→1946-78）

⑦ 異なる種類の2つの数字が続く場合，それらを区別しやすくするためによく綴りの短いほうをアルファベットで表わします。
　　　I have twelve 33-rpm records.
　　　（私は33回転レコードを12枚もっています）

◆基数

① 20までの基数と30, 40, 50, …, 90には，それぞれ個別の単語があります。
　　　three (3)／seventeen (17)／sixty (60)

② 上記を除く21～99は，10位と1位の単語をハイフンでつないで表記します。

thirty-two (32)／sixty-five (65)／ninety-eight (98)

③　100〜999は「100位の基数＋hundred [and] ＋下2桁の基数」と読み，また書きます。
　　　345────→three hundred [and] forty-five

④　4桁以上は1位から3桁ずつカンマで区切り（ただし，ちょうど4桁の基数はあまり区切らない），上の桁からmillion（100万），thousand（1000）などの単位をつけながら読みます。　　＊ただし，millionより大きい基数は米語と英語でちがいます。
　　　123,456→one hundred [and] twenty-three thousand, four hundred [and] fifty-six

⑤　1から20までの基数は単語で，21以上は数字で表記する傾向があります。また，7桁以上の概数は「数字＋（millionなどの単位の）単語」で書きます。
　　　one／two／three／four／five／・・・／twenty
　　　Let's count together from 50 to 100.
　　　I forgot whether it was 10 million or 100 million.

⑥　「基数＋名詞（単位以外）」の場合も上記⑤と同じです。
　　　I have seven dogs.
　　　There are 30 mothers in the park.
　　　I have made a deposit of 25 million yen.

◆序数

①　序数を表わすには原則として基数の語尾に-thをつけますが，20thまでと30th，40th，50th，・・・，90thにはそれぞれ個別の単語があります。

first (1st)／second (2nd)／third (3rd)／fourth (4th)／…／tenth (10th)／…／twentieth (20th)／thirtieth (30th)／fortieth (40th)／…／ninetieth (90th)

② 上記を除く21stから99thは10位の基数と1位の序数をハイフンでつないで表記します。
thirty-first (31st)／fifty-second (52nd)／sixty-eighth (68th)

③ 100th以上の序数は，まず基数として表わしはじめ，最後の単語だけ序数にします。
four hundred and fifty-second (452nd)／six hundredth (600th)

◆小数

小数はスペルアウトしません。整数部分が0（ゼロ）の場合，0は普通は書きますが，話すときには省略することもあります。

[正式]
① まず整数部分を基数として読み，小数点をandと読みます。つぎに小数部分は全桁を整数とみなして基数の読み方で読み，最後の桁の位を表わす単語，たとえば，tenthsやhundredths（序数に -sをつけた複数形）などをつけて読みます。なお，小数部分が全桁を整数とみなして1のときは -sは不要です。

8.5——→eight and five tenths
3.01——→three and one hundredth（-sはつけない）
12.07——→twelve and seven hundredths

② 整数部分が0のときは，小数点も読まずに小数部分だけを読みます。

   0.234──→two hundred [and] thirty-four thousandths
     ＊これは分数の読み方とまったく同じです。

   $\dfrac{234}{1000}$ ──→two hundred [and] thirty-four thousandths

[略式]
① 整数部分を基数として読み，小数点をpointと読み，小数部分は1桁ずつ数字を棒読みします。

   98.765──→ninety-eight point seven six five
   3.004──→three point 〈oh oh〉 four
   0.345──→[zero] point three four five

◆分数

分数は，分子・分母ともに1桁程度の単純な場合はスペルアウトすることがよくあります。

[正式]
① 日本語とは反対に，まず分子を基数として先に読み，つぎに分母を序数として読みます。スペルアウトするときは，分子と分母の間にハイフンをいれることもあります。分子が2以上なら，分母の序数には複数形の -sをつけます。また，分子が1の場合はoneではなく，aを用いる場合もあります。

   $\dfrac{1}{5}$──→one fifth (one-fifth, a fifth)
   $\dfrac{2}{6}$──→two sixths (two-sixths)

② 分母が2または4のときは，secondとfourthのかわりにhalfとquarterを使います。

$\dfrac{1}{2}$ ──→ one half (one-half, a half)
$\dfrac{1}{4}$ ──→ one quarter (one-quarter, a quarter)
$\dfrac{3}{4}$ ──→ three quarters (three-quarters)

③ 帯分数は，まず整数部分を基数として読み，つぎにandをつけて分数部分を読みます。

$3\dfrac{2}{3}$ ──→ three and two-thirds

[略式]
① まず分子を基数として読み，overをはさんで分母も基数で読みます。

$\dfrac{4}{7}$ ──→ four over seven

$5\dfrac{25}{48}$ ──→ five and twenty-five over forty-eight

◆数字と単位

① 単位は略記を用いる場合とスペルアウトする場合があります。略記にはピリオドも複数形の-sもつけません。

78 km／78 kilometers
34 mg／34 milligrams

② 略記を用いる場合，数字と単位の間は原則ワン・スペースあけます。ただし，％，℃，□°はあけません。

250 km／72 cm／2.8 ℓ／5 kg
24％／12℃／360°

チャレンジ・テストの解答

チャレンジ・テスト① 37ページ

① 6 (six)
② 13 (thirteen)
③ Twenty three is less than twenty four. [23は24より少ないです。]
④ Yellow [黄色] ◇The 2nd (second) from the right is yellow. [右から2番目は黄色です。]
⑤ Saturday [土曜日] ◇Four days from Tuesday is Saturday. [火曜日から4日先は土曜日です。]
⑥ 4 (four)
⑦ 3 (three) [3桁] ◇It is a three-digit number. [それは3桁の数です。]
⑧ an even number [偶数]
⑨ an odd number ／ an even number [奇数／偶数]
⑩ 12 (twelve), 24 (twenty-four), and 36 (thirty-six)
⑪ 1 (one) , 2 (two), 3 (three), 4 (four), 6 (six),
8 (eight), 12 (twelve) , and 24 (twenty-four).
⑫ 20 cm (twenty centimeters)
⑬ 2 (two) people, and 4 (four) people ／ 4 (four) people
[2人, 4人 ／ 4人]
⑭ A:2,350 (two thousand three hundred fifty),
B:2,449 (two thousand four hundred forty-nine)
⑮ 2000 (two thousand)
⑯ rounded [off]: 3,600 (three thousand six hundred) people ／
rounded up: 3,600 (three thousand six hundred) people ／
rounded down: 3,500 (three thousand five hundred) people [四捨五入3600人／切り上げ3600人／切り捨て3500人]

チャレンジ・テストの解答　187

⑰ 3＜5＜7＜9＜12 (three, less than five, less than seven, less than nine, less than twelve)

⑱ －14 (Negative fourteen) is larger.

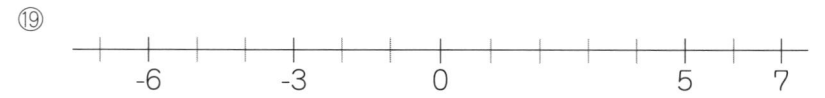

⑲

```
—+—+—+—+—+—+—+—+—+—+—+—+—+—
   -6      -3         0           5   7
```

⑳ The difference in temperature is 7 (seven) degrees. 〔温度の差は7℃です。〕

㉑ 8.4 (eight point four)

㉒ 0.4 (zero point four)

㉓ three point four seven nine

㉔ $5\frac{2}{3}$ (five and two thirds)

㉕ $\frac{3}{11}$ (three elevenths)

㉖ A 0.5 (zero point five), $\frac{1}{2}$ (one over two)　B 1.75 (one point seven five), $1\frac{3}{4}$ (one and three over four)　C 2.2 (two point two), $2\frac{1}{5}$ (two and one over five)

㉗ $\frac{7}{4}=1.75$ (one point seven five), $\frac{11}{5}=2.2$ (two point two), $\frac{5}{2}=2.5$ (two point five)

チャレンジ・テスト② 66ページ

① 3＋7＝10, 10 (ten)

② 21＋35＝56, 56 (fifty-six)

③ 10×4＋1×8＝48, 48 (forty-eight)

④ 3.6＋2.8＝6.4, 6.4 m (six point four meters)

⑤ 67－34＝33, 33 (thirty-three)

⑥ There are 9 (nine) more red cards. 〔赤いカードが9枚多いです。〕

⑦ 46−28=18，18 children［18人］◇18 (Eighteen) children remain.［残っているのは18人です。］

⑧ 8−3=5，5ℓ［5リットル］◇5ℓ (Five liters) remain.［残りは5ℓです。］

⑨ Mr. Takahashi is 9 kg (nine kilograms) heavier.
［高橋君が9kg重い。］

⑩ two times seven ／ four times eight ／ fifteen times eleven ／ twenty-three times thirty-two ／ forty-two times one hundred and eight

⑪ 12×4=48，48 months ◇There are 48 (forty-eight) months in 4 (four) years.［4年は48か月です。］

⑫ 3×4=12，12 oranges ◇There are 12 (twelve) oranges.［みかんは12個あります。］

⑬ 3 hours 30 minutes［3時間30分］◇He jogs 3 (three) hours 30 (thirty) minutes in a week.［1週間では3時間30分になります。］

⑭ 7×9=63，63 people［63人］◇63 (Sixty-three) people did.［63人でした。］

⑮ 102÷3=34，34 (thirty-four)

⑯ 9.2÷2.3=4，4 (four)

⑰ 26÷4=6 …2，6［6個］／ 2 (two)［2個］◇Each person gets 6 (six) peaches per person, and 2 (two) are left over.［1人あたり6個ずつで，余りは2個です。］

⑱ (例) There are 18 (eighteen) flowers. How many people can share them when 3 (three) flowers are given to each person?［18本の花があります。1人に3本ずつ分けると，何人に分けられるでしょうか。］

⑲ 53 (fifty-three)，506 (one hundred and six)

⑳ (850−65)×5=3,925，3,925 yen［3925円］◇It costs ¥3,925 (three thousand nine hundred and twenty-five yen) in

㉑ 1000×3−(250×3+350×4)=850, ¥850 [850円] ◇The change is ¥850 (eight hundred and fifty yen). [おつりは850円です。]
㉒ x＝(50−8)÷7=6, 6 (six)
㉓ 8x×6=144, x=3 ◇x is 3 (three). [xは3です。]
㉔ 3x÷5=9, x=15 ℓ ◇1 (One) bottle of juice is 15 ℓ (fifteen liters). [ジュース1本は15 ℓです。]

チャレンジ・テスト③　88ページ

① 2 (two) minutes 4 (four) seconds [2分4秒]
② 9:05 (nine 〈ou〉 five) a.m. [午前9時5分]
③ 2 hours 10 minutes [2時間10分] ◇He played for 2 (two) hours 10 (ten) minutes. [遊んだ時間は2時間10分です。]
④ 8 o'clock 30 minutes [8時30分] ◇We will arrive at the station at 8:30 (eight thirty). [駅に着くのは8時30分です。]
⑤ 21+46=67, 67 cm ◇We get 67 cm (sixty-seven centimeters). [それは67cmです。]
⑥ 403−400=3, 403 cm (Four hundred and three centimeters) is 3 cm (three centimeters) longer than 4 m (four meters). [403cmのほうが4mより3cm長いです。]
⑦ 460+280+340=1,080, 1,080 m ◇It is 1,080 m (one thousand and eighty meters) from the house to the station. [家から駅までは1080mです。]
⑧ 75×3+200=425, 4 m 25 cm ◇We get 4 m 25 cm (four meters twenty five centimeters). [ヒモの長さは4m25cmです。]
⑨ 68 $d\ell$ ◇There are 68 $d\ell$ (sixty-eight deciliters). [それは68 $d\ell$です。]

⑩ 7 ℓ+3 dℓ=73 dℓ，73 dℓ (deciliters) ／ 7.3 ℓ (liters) ◇We get 73 dℓ (seventy-three deciliters). There are 7.3 ℓ (seven point three liters). [それは73 dℓです。また，7.3 ℓです。]

⑪ 500×7=3500，3.5 ℓ ◇There are 3.5 ℓ (three point five liters). [それは3.5 ℓです。]

⑫ 2.8−0.9=1.9，1.9 ℓ ◇1.9 ℓ (One point nine liters) spilled out. [こぼしたのは1.9 ℓです。]

⑬ 48÷4=12，12 ℓ ◇Each will get 12 ℓ (twelve liters). [1個あたり12 ℓになります。]

⑭ 3,052 g ◇It is 3,052 g (three thousand fifty-two grams). [それは3052 gです。]

⑮ 53−46=7，7 kg ◇The elder brother is 7 kg (seven kilograms) heavier. [兄のほうが7kg重いです。]

⑯ 4kg34 g＜4056 g＜4kg87 g (Four kilograms thirty-four grams, less than four thousand fifty-six grams, less than four kilograms eighty-seven grams)

⑰ 1032−230=802，802 g ◇The salt weighs 802 g (eight hundred and two grams). [塩の重さは802 gです。]

⑱ (78+82+85+91+94)÷5=86，86 g ◇It is 86 g (eighty-six grams). [りんご1個あたり86 gです。]

⑲ 144÷12=12，12 flowers [12本] ◇There are 12 (twelve) flowers per square meter. [1m² あたり12本です。]

⑳ 360÷3=120，750÷5=150，The cloth priced at ¥360 (three hundred sixty yen) for 3 m (three meters) is ¥20 (twenty yen) cheaper per meter. [3m360円の布のほうが1mあたり20円安い。]

㉑ 255÷3=85，85 km ◇The speed of the car was 85 km (eighty-five kilometers) per hour. [自動車の時速は85kmです。]

チャレンジ・テスト④　126ページ

① Line A and line C are parallel to each other, and line B and line E are perpendicular to each other. [AとCが平行，BとEが垂直]

② a is 45° (forty-five degrees), b is 135° (one hundred and thirty-five degrees), c is 45° (forty-five degrees), and d is 135° (one hundred and thirty-five degrees). [aは45°，bは135°，cは45°，dは135°]

③ a is 40° (forty degrees), b is 140° (one hundred and forty degrees), c is 60° (sixty degrees), d is 270° (two hundred and seventy degrees), and e is 70° (seventy degrees). [aは40°，bは140°，cは60°，dは270°，eは70°]

④ a is 135° (one hundred and thirty-five degrees), b is 270° (two hundred and seventy degrees), and c is 60° (sixty degrees). [aは135°，bは270°，cは60°]

⑤ $\angle A = 180° - (\angle B + \angle C) = 180° - (50° + 40°) = 90°$ $\angle a = \frac{1}{2}\angle A = 90° \div 2 = 45°$, 45°◇Angle a is 45 (forty-five) degrees. [∠aは45°です。]

⑥ AC=AD+DC=AB+BD=4+4=8，8cm◇AC is 8 cm (eight centimeters) long. [ACは8cmです。]

⑦ trapezoids, squares, rectangles, parallelograms, rhombuses [台形，正方形，長方形，平行4辺形，ひし形]

⑧ ∠a＝115° (one hundred and fifteen degrees)，∠b＝65° (sixty-five degrees)，∠c＝150° (one hundred and fifty degrees)，∠d＝140° (one hundred and forty degrees)

⑨ 360°÷5=72°，72°。9 [9本] ◇The central angle of regular pentagon is 72° (seventy-two degrees). It has 9 (nine) diagonals. [正5角形の中心角は72°です。対角線は9本です。]

⑩ square, rectangle, rhombus, parallelogram [正方形，長方形，ひし形，平行4辺形]

⑪ ∠e＝50° (fifty degrees)，∠f＝110° (one hundred ten degrees)，

∠g＝70°(seventy degrees)，∠h＝130°(one hundred thirty degrees)

⑫ A－point symmetry［点対称］，B－line symmetry［線対称］，C－point symmetry［点対称］，line symmetry［線対称］，D－line symmetry［線対称］

⑬

⑭ 2×（3＋5）＝16，16 cm◇The diameter of circle C is 16 cm (sixteen centimeters).［円Cの直径は16cmです。］

⑮ 2×π×7＝43.96，43.96 m◇The circumference of the pond is 43.96 m (forty-three point nine six meters).［池の周囲は43.96mです。］

⑯ Two rectangles with side-lengths 3 cm (three centimeters) and 9 cm (nine centimeters), two rectangles with 3 cm (three centimeters) and 5 cm (five centimeters), and two rectangles with 5 cm (five centimeters) and 9 cm (nine centimeters)［たて3cmとよこ8cmの紙が2枚，3cmと5cmのが2枚，5cmと8cmのが2枚］

⑰ (1)point d and point h　(2)edge de　(3)face D［(1) 点dと点h　(2) 辺de　(3) 面D］

⑱ (A)triangular prism［3角柱］　(B)triangular pyramid［3角錐］　(C)hemisphere［半球］

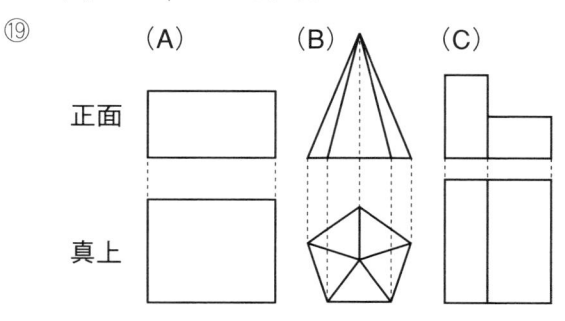

⑲

チャレンジ・テスト⑤　150ページ

① 48÷4=12, 12×12=144, 144 cm² ◇The area of the square is 144 cm² (one hundred forty-four square centimeters). ［正方形の面積は144cm²です。］

② 20×40+25×15=1175, 1175 m² ◇The area is 1,175 m² (one thousand one hundred seventy-five square meters). ［その面積は1175m²です。］

③ 7×9×6−3×4×6=306, 306 (three hundred and six) m³ ◇Its volume is 306 m³ (three hundred and six cubic meters). ［その体積は306m³です。］

④ (20−2×2)×(24−2×2)×(17−2×1)=4800, 4800 cm³ =4.8 ℓ, 4.8 ℓ ◇Up to 4.8 ℓ (four point eight liters) of water can be stored in the tank. ［水槽の最大の水の量は4.8ℓです。］

⑤ (A) $\frac{1}{2}$ (6×6) =18, 18 cm² (eighteen square centimeters), (B) 4×6=24, 24 (twenty-four) cm², (C) $\frac{1}{2}$ (5×4) + $\frac{1}{2}$ (5×2) =15　15 (fifteen) cm², (D) $\frac{1}{2}$ (6+3) ×3+ $\frac{1}{2}$ (6+3) ×3=27　27 (twenty-seven) cm²

⑥ (A) 表面積 $\frac{1}{2}$ (3×4)×2+5×7+3×7+4×7=96, 96 cm² (ninety-six square centimeters)／体積 $\frac{1}{2}$ (3×4)×7=42, 42 (forty-two) cm³　(B) 表面積 2×3.14×3×8+3.14×3×3×2=207.24, 207.24 (two hundred seven point two four) cm²／体積 3.14×3×3×8=226.08, 226.08 (two hundred twenty-six point zero eight) cm³　(C) 表面積 5×5×2+5×8×4=210, 210 (two hundred ten) cm²／体積 5×5×8=200, 200 (two hundred) cm³

⑦ 3.14×8×8=200.96, 200.96 cm² ◇The area of the circle is 200.96 cm² (two hundred point nine six square centimeters).

[円の面積は200.96 cm²です。]

⑧ (A) $\frac{1}{3}$ (4×4×12) =64, 64 cm³ (sixty-four square centimeters)
(B) $\frac{1}{3}$ (3.14×3×3×12) =113.04, 113.04 (one hundred thirteen point zero four) cm³   (C) $\frac{4}{3}$ (3.14×3×3×3) = 113.04, 113.04 (one hundred thirteen point zero four) cm³

チャレンジ・テスト⑥　161ページ

①

year	people
1	14
2	23
3	18
4	25
5	17
6	20

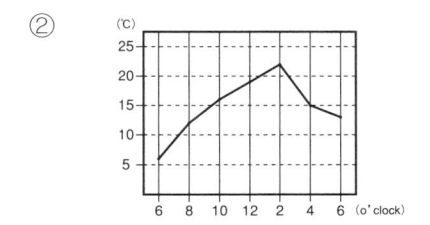

②

③ (80−48)÷80×100＝40, 40%◇40 (Forty) percent are boys. [それは40パーセントです。]

④ 24÷150×100＝16, 30÷150×100＝20, 12÷150×100＝8, 48÷150×100＝32, 36÷150×100＝24, Red is 16% (sixteen percent), blue is 20% (twenty percent), purple is 8% (eight percent), yellow is 32% (thirty-two percent), and green is 24% (twenty-four percent). [赤が16％, 青が20％, 紫が8％, 黄色が32％, 緑が24％]

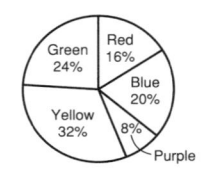

	Purple			
Red 16%	Blue 20%	8%	Yellow 32%	Green 24%

⑤ (5+10+0+7+13+1)÷6=36÷6=6　An average of 6 (six) students per day forgot something. The total number of students for the week is 36 (thirty-six). [1日の平均は6人。1週間の延べ人数は36人です。]

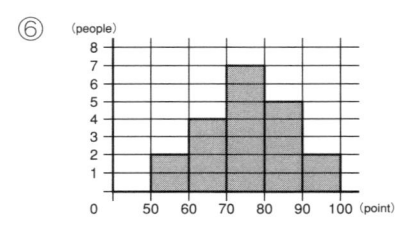

⑥

チャレンジ・テスト⑦　177ページ

① ① 9：8 (nine to eight) ② 3：4 (three to four) ③ 6：5 (six to five)

② 50：□=2：5　□=50×5÷2=125，125 ㎖◇125 ㎖ (One hundred twenty-five milliliters) of hot water is needed. [お湯は125㎖必要です。]

③ 3：2=150：x　x=2×150÷3=100，100 cm=1 m◇Taro's shadow is 1 m (one meter) long. [太郎君の影の長さは1mです。]

④ (4×3)×(12×3)÷2=216，216 cm² ／ (4×$\frac{1}{2}$)×(12×$\frac{1}{2}$)÷2=6，6 (six)　cm²◇The area of the enlarged figure is 216 c㎡ (two hundred sixteen square centimeters). [その拡大図の面積は216 cm²です。] ／ The area of the reduced figure is 6 cm² (six square centimeters). [その縮小図の面積は6 cm²です。]

⑤ (1×200)×(1.5×200)+(3×200)×(2.5×200)（cm²）＝2×3+6×5（m²）＝36（m²），36 m²◇The area of the garden is 36 m² (thirty-six square meters).［菜園の面積は36m²です。］

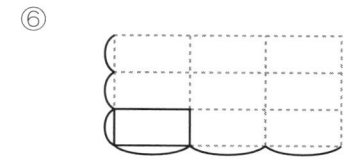

⑥ 9 (Nine) times.［9倍］

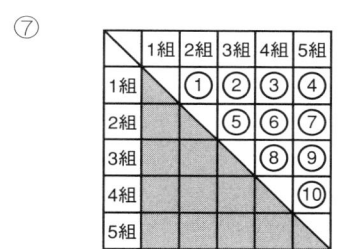

⑦ 10 (Ten) combinations.［10通り］

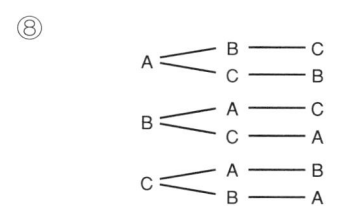

⑧ 6 (Six) cases.［6通り］

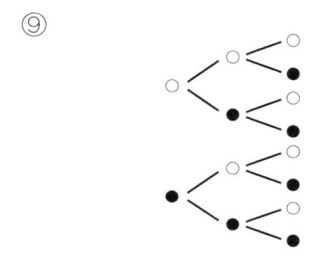

⑨ 8 (Eight) cases.［8通り］

●著者紹介

銀林　浩（ぎんばやし・こう）
▶1927年，東京に生まれる。東京大学理学部数学科を卒業。
▶明治大学教授を経て，同大学名誉教授。専門は代数学・整数論・経営数学・数学教育。
▶元数学教育協議会委員長。現在，同会常任委員。ルンビニー学園幼稚園講師。
日教組教研数学教育分科会共同研究者。数学教育研究会顧問。
▶おもな著書：『水道方式による計算体系』『人間行動からみた数学』（明治図書），
『数の科学』『量の世界』（むぎ書房），『人文的数学のすすめ』（日本評論社），
『数・数式と図形の英語』『数量と単位の英語』（日興企画），
『子どもはどこでつまずくか』（国土社）ほか多数。

銀林　純（ぎんばやし・じゅん）
▶1959年，東京に生まれる。東京大学理学部数学科を卒業。同大学の博士課程を中退。
▶1986年，富士通株式会社に入社。以来，コンピュータの業務応用システムを
開発する技術の研究と適用・推進に従事した。
▶1991年より94年までオクスフォード大学に大学院生として社費留学し，博士号を取得。
▶おもに情報処理学会と人工知能学会にてコンピュータ関連の論文を発表。
▶ISO，OMG，JISといった標準化・規格化の活動にも参画し，
とくに業務モデリング技術の整備と普及に携わった。
▶おもな著書：『数・数式と図形の英語』『数量と単位の英語』（日興企画）

図解 子供にも教えたい 算数の英語 ―豊富な用語と用例	
2006年3月9日…初版発行	発行者…竹尾直文
2021年11月20日… 3版発行	制作者…友兼清治
	発行所…株式会社 日興企画
	〒104-0032
	東京都中央区八丁堀4-11-10
	電話　03-6262-8127
	FAX　03-6262-8126
著者……銀林　浩	印刷所…シナノ印刷株式会社
銀林　純	定価……カバーに表示してあります。
ISBN4-88877-643-1 C2082	ⓒKo GINBAYASHI & Jun GINBAYASHI 2006, Printd in Japan

【小社出版物のご案内】定価・価格はすべて税込みです。

●篠田義明の実用英語シリーズ　平均220ページ・A5判

国際会議・スピーチに必要な英語表現
出迎え・就任・乾杯・哀悼などの挨拶－開会・議事進行・閉会など司会や議長の言葉－提案・質議など会議中の用語。　★2970円(本体2700円+税10%)

CD版 別売
3630円
(本体3300円+税10%)

科学技術論文・報告書の書き方と英語表現
目次・序論・研究方法など論文や報告書の書き方と表現。目的・実験・試験・調査・例題など論文の展開に役立つ表現。　★3300円(本体3000円+税10%)

科学技術論文に頻出する英語表現 1　●数式・図形・測定・分析＝編
加減乗除・比例－方程式・微分積分・集合・確率－平面図形・立体図形・角度・面積・体積－測定・観察・試料・分析。　★2310円(本体2100円+税10%)

ネゴシエーション・会議に必要な英語表現
意見や感想を述べる－意向を問う－提案や報告を検討する－議論や質疑を展開させる－会話中によく用いることば。　★2750円(本体2500円+税10%)

数理・理工英語の基本用語と活用文例
基本用語を五十音順に配列し、対応する基本英文をできる限り完全な文の形で紹介した和英対照の表現事例集。　★3630円(本体3300円+税10%)

コンピューターとインターネット英語の用語と文例
ワープロ・表計算・データベース・グラフィックスなどパソコンソフトからインターネット・メールまでの基本用語。　★3080円(本体2800円+税10%)

パーティー・プレゼンテーションに必要な英語表現
自己紹介・就退任・表彰・創立記念・慶弔儀式－経営や営業の方針・新商品紹介・販促・調査報告－発言中にはさむ言葉。　★2750円(本体2500円+税10%)

CD版 別売
3630円
(本体3300円+税10%)

●富井篤の[わかる・使える]実務英語シリーズ　平均270ページ・A5判

第1巻　新・実務英語入門 書き方と訳し方
表現・文法・構文から英文ルール・記号まで、その重要項目を厳選し、初心者向けに解説したダイジェスト版。　★3080円(本体2800円+税10%)

第3巻　数量英語の活用文例集
数量表現の基本パターン－厚さ－距離－高さ－長さ－時間－年代－圧力－重さ－比重－トルク－温度－加速度。　★2750円(本体2500円+税10%)

第4巻　実務英語の簡潔表現と文例集
英文を書くときに必要になる主要な表現77件を選び、さらに300の細項目にわけて内容ごとにその書き方を解説。　★4180円(本体3800円+税10%)